A. _____ _____ FIRES

Fires

A M E R I C A ' S

A Historical Context for Policy and Practice

STEPHEN J. PYNE

THE
FOREST
HISTORY
SOCIETY

ISSUES
SERIES

Forest History Society
Durham, North Carolina

FOREST HISTORY SOCIETY ISSUES SERIES

The Forest History Society was founded in 1946. Since that time, the Society, through its research, reference, and publications programs, has advanced forest and conservation history scholarship. At the same time, it has translated that scholarship into formats useful for people with policy and management responsibilities. For seven decades the Society has worked to demonstrate history's significant utility.

The Forest History Society's Issues Series is one of the Society's most explicit contributions to history's utility. The Society selects issues of importance today that also have significant historical dimensions. Then we invite authors of demonstrated knowledge to examine an issue and synthesize its substantial literature, while keeping the general reader in mind.

The final and most important step is making these authoritative overviews available. Toward that end, an initial distribution is made to people with education, management, policy, or legislative responsibilities who will benefit from a deepened understanding of how a particular issue began and evolved. The books are commonly used in education programs throughout North America and beyond.

The Issues Series—like its Forest History Society sponsor—is nonadvocatory and aims to present a balanced rendition of often contentious issues.

Other Issues Series titles available from the Forest History Society:

American Forests: A History of Resiliency and Recovery
Newsprint: Canadian Supply and American Demand
Forest Pharmacy: Medicinal Plants in American Forests
Forest Sustainability: The History, the Challenge, the Promise
Canada's Forests: A History
Genetically Modified Forests: From Stone Age to Modern Biotechnology
American Forested Wetlands: From Wasteland to Valued Resource

To Sonja
who learned to adapt to fire

THE FOREST HISTORY SOCIETY

The Forest History Society is a nonprofit educational and research institution dedicated to the advancement of historical understanding of human interaction with the forest environment. It was established in 1946. Interpretations and conclusions in FHS publications are those of the authors; the institution takes responsibility for the selection of topics, competency of the authors, and their freedom of inquiry.

This revised edition was published with support from David L. Luke III, The National Forest Foundation, MeadWestvaco Corporation, and the Lynn W. Day Endowment for Forest History Publications.

Printed in the United States of America

Forest History Society
701 William Vickers Avenue
Durham, North Carolina 27701
(919) 682-9319
www.foresthistory.org

SD
421.3
.P958
2009

©2010 by the Forest History Society

Revised edition

Design by Zubigraphics, Inc.

On the cover: Residents flee their hillside homes on Los Angeles' San Fernando Valley during the Sesnon Fire in October 2008. Ignited by a power line blown down by high winds, the fire consumed nearly 15,000 acreas in a heavily populated area. (AP photo/Dan Steinberg)

Library of Congress Cataloging-in-Publication Data

Pyne, Stephen J., 1949–
    America's fires : a historical context for policy and practice / Stephen J. Pyne. — Rev. ed.
        p. cm.
    Includes bibliographical references.
    Summary: "*America's Fires* reviews the historical context of our fire issues and policies that can inform the current and future debate. The forecast makes it imperative that the nation review its policies toward wildland fires and find ways to live with them more intelligently"— Provided by publisher.
    ISBN 978-0-89030-073-2 (pbk. : alk. paper)
    1. Wildfires—United States—History. 2. Forest fires—United States—History. 3. Fire management—United States—History. 4. Forests and forestry—Fire management—United States—History. I. Title.
    SD421.3.P958 2009
    363.37'9—dc22
                                                        2009043660

7222410

CONTENTS

LIST OF FIGURES

Frontispiece: Started by lightning on the Seminole State Forest in central Florida on May 7, 2007, the Lee Fire eventually burned about 2,600 acres of mostly Southern rough (pines, gallberry, and palmetto). The state's division of forestry needed air tankers and helicopters to prevent the fire from consuming a small residential development just outside the state forest.

FOREWORD

I n 1997 the Forest History Society and Stephen J. Pyne published *America's Fires: Management on Wildlands and Forests* as an early edition in the Society's Issue Series. At the time, the United States had completed what Pyne characterized as a "tumultuous decade" of wildfires, and it appeared the nation would address its policy on fire and forests.

An overhaul of forest fire policy never occurred, and the following decade was even more tumultuous. Wildfires continued unabated in number, size, and impact on federal land in the West; exacerbated by warming trends, continued fuel accumulation in forest understories, and the expansion of homebuilding in forested areas. As of this printing, the Forest Service is spending nearly half of its annual budget on fire suppression, and that portion increased every year during the last decade. In the process, the capacity to manage the national forests for all of their values diminishes steadily.

In the South, factors that sustained low levels of wildfires on private forestland have undergone massive change. The almost complete transfer of forest ownership from forest products firms to financial investors has been accompanied by the collapse of the South's industrial fire suppression capacity. Public forestry agencies have endured budget cuts and diversion of resources from fire control into regulatory enforcement. The number of prescribed burns has fallen because of costs, liability concerns, insufficient staff, changes in rural culture, and air quality regulations. A return to large, uncontrolled fires has begun.

Hovering over all of this is an immense change in U.S. demographics. In 2007 the nation counted 300 million people. In another 30 to 40 years the country is forecast to see a population increase to 400 million, with most of the additional people living in the West and South in an increasingly urbanized society. That forecast makes it imperative that the nation review its policies toward wildland fires and find ways to live with them more intelligently.

In 2007 the National Coalition of Prescribed Fire Councils was formed to steer a much more focused effort to restore the use of fire in fire-dependent forest ecosystems. In 2008 the National Commission on Science for Sustainable Forestry (NCSSF) completed six years of intensive work to assess the scientific basis of sustainable forestry and biodiversity conservation in the United States. Scientists in forestry and forest ecology have gained important new understanding of the role of fire in forest ecosystems, the changes in those ecosystems from their historical conditions, and the need to continue to use fire to sustain ecosystem health. NCSSF concluded that the limitation to sustaining America's forests was not science but the incoherence of forest policies, laws, regulations, and court rulings that have accumulated through recent decades. NCSSF has called for a presidential commission to examine America's forest policies and create a more modern, scientifically sound, and realistic framework of policies and laws for the future. NCSSF has also called for leaders from diverse interests to steer such a process, the formation of a forestry caucus in Congress, and the establishment of strong forestry groups within the nation's governors' associations.

It is in that context that the Forest History Society and Stephen Pyne are publishing this revised Issue Series book on fire and America's forests and wildlands. If policies are to be reviewed and updated, it is important to understand the historical role of fire in forests and the origins and evolution of our existing policies and approaches. As Pyne emphasizes, intelligent changes to American policies and programs related to wildland fire cannot occur independently from intelligent changes in approaches to managing those forests. There is no simple "fix" to the problem.

The Issue Series books—like the Forest History Society—are nonadvocacy. The series aims to present a balanced rendition of information about topical and often contentious issues in a factual, historical context. We believe this revised edition is timely in light of the continued challenges posed by wildland fire at a time when more and more Americans are concerned about retaining forestland as forest in the face of relentless urbanization. The information in this publication will be important and useful to anyone who is involved in reshaping the policies that address America's forests and their relationship to fire.

R. Scott Wallinger and Steven Anderson

AUTHOR'S NOTE ON THE NEW EDITION

W hen *America's Fires: Management on Wildlands and Forests* was published in early 1997, America's fire establishment had completed what it regarded as a tumultuous decade. Drought and past land use had combined to unleash two summers that most observers at the time considered epochal. The 1988 fire season, best known for the long summer firefight at Yellowstone National Park, carried fire issues to the public, advertising the ecological significance of free-burning fire and the dilemmas of trying to manage it. The 1994 season did something similar for the fire community, reinforcing the true costs and complexities of fire's management. Together, those years appeared as a kind of concluding chapter to a century of state-sponsored fire protection. *America's Fires* reflected that understanding.

In fact, those seasons were less an end than a rekindling. The decade that followed witnessed, for many western states, the largest fires on record. The media became sensitized to fire as news. Efforts to reform our approach to fire management inspired national programs but also led to local lapses. Drought continued, the legacy of past practices became more undeniable, nature proved uncooperative, and politics tricky. The national saga of fire was measured not by years passed but by events experienced, and as each new fire season piled new incidents upon old, *America's Fires* became less useful as a guide to the contemporary scene. The discussion was no longer dominated by the ancient quarrel between fighting fires and lighting them; the problem was how to do both.

The Forest History Society believed the wisest solution was to present a new edition, one in a simpler narrative style that would allow the story to unfold more naturally, in what might be regarded as a kind of literary version of adaptive management. This new edition is thus a complete revision, rewritten root and branch. Some figures, updated, remain, and a few passages here and there have survived. But in its organization, its arguments, and its prose, this is a fresh text, and its new subtitle, *A Historical Context for Policy and Practice*, reflects that fact. What careful readers might find, however, is that phrases and concepts I have developed over the course of 30 years of writing and speaking about fire have reappeared.

The truth is, I'm not endlessly inventive. There are only so many ways to say concisely what I want. When asked to write an extended, popular essay that would give expression to my basic understanding of fire history, I have had to repair, reuse, renew, and recycle phrases from several thousand pages of previous writing and decades of lecturing, so a lot of fragments from other works, like so much flotsam and jetsam, have been gathered and assembled into new context. Where I have picked up those pieces, I have tried to present them in ways that convey my current state of understanding, much of which is scattered across many books, to an audience that those books might not otherwise reach. If others had cited those works, I could quote them without hesitation; but it seems indecent to quote oneself. For a fuller and more lively exposition of the principal themes, I recommend in particular *Tending Fire: Coping with America's Wildland Fires*.

Special thanks to Steven Anderson of the Forest History Society for prodding me to rewrite this study; to David L. Luke III and the National Forest Foundation for funding; to Scott Wallinger, Zachary Prusak, and Jim Brenner for information; to Sally Atwater for careful editing; and to all those who over the years have enriched my understanding of America's fire scene.

F ire has existed on Earth since lightning first struck terrestrial plants amid an oxygenated atmosphere some 400 million years ago. Early hominids captured fire perhaps 1.5 million years ago, and humans have enjoyed a monopoly over fire's manipulation ever since. Free-burning fire has been a constant ecological presence on the American landscape.

But its expression has changed, often dramatically, as new climates, new peoples, and new land uses have become predominant. The evolution of an industrial society committed to the controlled combustion of fossil fuels fundamentally restructured the continent's fire regimes, and fire prevention and suppression in wildlands became a policy goal. Recent efforts to reintroduce fire into ecosystems have created new challenges. Since the problems look the way they do in good part because of how they have developed historically, this brief history can offer insights for future discussion and decision making.

- Fire is, in its origins, a biological phenomenon because it feeds on plant matter; it both affects ecosystems and is altered by them. But fire is also a cultural phenomenon because humans exercise a unique monopoly over its manipulation.

- Humans have spread fire on the landscape to make their world more habitable. They have exercised their firepower in stages. Aboriginal fire for foraging, hunting, and fishing relied on control over ignition alone. Agricultural fire added control over fuels as well.

- In what became the United States, European settlers used fire to clear land permanently and maintain it for farming and herding, but where they also fixed landownership, fire became destructive and had to be contained within certain places and times.

- Industrialization added another level of human manipulation: people burned fossil biomass, not simply surface biomass, and relied on machinery to project their firepower and to alter the character of fire on the landscape.

- In the late 19th century, damaging fires on the frontier prompted conservationists to campaign for forest reserves that could be protected from both fire and axe. Fire protection was assumed to be both necessary and possible, and soon a national effort led by foresters was underway.

- Early in the 20th century, huge fires in the West tested the ideas behind fire control. Some people promoted a strategy of frequent light burning to prevent large fires, but a model of fire suppression and fire exclusion won favor. The U.S. Forest Service eventually dominated every aspect of wildland fire management.
- During the Depression, the Civilian Conservation Corps provided the manpower to extend fire protection into the backcountry. Subsequently, World War II blurred the line between firefighting and civil defense, and after the war, surplus equipment was mobilized for a military style of fire suppression. Only in the Southeast was controlled burning restored to reduce fuels in the forest.
- The early 1960s saw a change in attitudes. Fire ecologists presented evidence that fire belonged in the landscape. Many citizens were becoming skeptical of government experts, including foresters, and the Wilderness Act provided a legal basis for challenging the federal land management policies.
- The policy of fire exclusion persisted, however, until its economic cost and ecological legacy demanded a reintroduction of fire. By the mid-1970s, federal agencies had pulled back from the fire suppression model and embraced a mixture of fire practices, especially forms of prescribed burning.
- Forest conditions had changed over the decades, however, and after several years of drought, in 1988 the West experienced large, uncontrolled fires, which have continued. In the 1990s and early 2000s, violent fires spread into the exurbs, destroying homes, killing firefighters, and straining Forest Service budgets. The new approaches were proving difficult to implement.
- In 2008 "appropriate management response" became federal policy: fires could be prescribed, suppressed, or allowed to burn, depending on the land, the resources, and the risks.
- It is clear that America does not have a fire problem: it has many fire problems, each requiring particular, distinctive responses. The dominant focus, however, remains fire on public lands, which is where most large fires reside.
- On such fire-prone public lands, we can consider mixing and matching four approaches, each of which has its advantages and disadvantages, both environmental and political:
  1. Leave fire management to nature as much as possible, particularly through the adoption of wildland fire use.
  2. Exclude fire through prevention and suppression, although at the risk of creating bigger fires when those measures fail.
  3. Do the necessary burning ourselves through varieties of prescribed fire.
  4. Redesign the landscape so that fires of any sort will behave as desired.
- Fire research still needs to address fire in its full biological identity and integrate humans into its agenda—because people are the most powerful agents in fire's presence in the American landscape.

There was a time when fire did not exist on Earth. But that epoch ended with the advent of life. Marine life pumped the atmosphere full of oxygen, terrestrial life lathered the lands with hydrocarbons, and lightning supplied the necessary spark to kindle their reaction. Even the chemistry of that reaction is a *bio*chemistry: fire takes apart what photosynthesis puts together. When this occurs in cells, it is called respiration, and when it happens in the wide world, we call it fire. By the early Devonian period, 420 million years ago, open combustion was leaving charcoal evidence in the geologic record. Earth has been a fire planet ever since.

Yet fire is rarely understood as a biological phenomenon. Because we can break combustion down into its chemical components and replicate that process in fireplaces or engines and because free-burning fires happen amid dry spells and high winds, we think of fire simply as a chemical reaction shaped by its physical circumstances. This can be true in labs and factories, and when fire is free-burning over landscapes. But in the wide world, fire originates in biology. Unlike ice storms, floods, debris flows, and winds, all of which can occur without a particle of life present, fire cannot exist, much less spread, apart from its biological matrix. Unlike those purely physical events, fire does not act *on* but spreads *through*—feeds on—that medium. Environmental factors matter only insofar as the living world integrates them. Fundamentally, fire is biologically constructed.

Fire and life thus shape each other. Fire can influence organic evolution overall, and it helps direct the ways many ecosystems work. Equally, anything that rearranges the living world can alter drastically the kinds of fire that occur. Browsers, grazers, decomposers—fire potentially competes with all of them for a common crop of plant matter. What is food for one is fuel for the other, and it will be oxidized by combustion that is either slow (metabolism) or fast (fire). As life has evolved, so has fire, and as the living world has morphed endlessly into new arrangements, so has fire. The living world can shape fire in ways it cannot influence floods or winds.

A peculiar pairing of wetting and drying underwrites the basics of fire's ecology: a land must be wet enough to grow combustibles and dry enough to ready them for burning. Places that are always wet or always dry—rainforests or stony deserts—do not burn. Normally wet places burn during droughts, and normally dry places burn after the odd rain. If a place experiences annual dry seasons, it can burn annually; if it is subject to periodic drought and deluge, it can burn periodically; if the mix of wetting and drying occurs on the order of centuries, then it burns accordingly.

Still, one part of fire's combustion triangle remained outside the control of life. Combustibles cannot cause fire by themselves; they need a spark, and life had no routine way to supply it. Ignition depended on a physical spark, which might come from rolling stones, volcanic eruptions, even at times spontaneous combustion, but usually from lightning. This kindler of fire remains common today, and in many landscapes, it is dominant.

Lightning must act on a suitable setting; the geography of lightning bears little relationship to the geography of fire for the simple reason that concentrated lightning comes from thunderstorms that drown most ignitions. Where lightning fire predominates, it does so by coming after a long dry season or in mountains where thunderstorm downwashes become separated from lightning strikes or in arid landscapes where rain evaporates before it wets the ground. The resulting combination of dry lightning and dry wind is ideal for starting and spreading fire, but it also inscribes a patterning of fire that is lumpy in space and time. There are places and times that burn routinely, and others that burn hardly ever. In the Paleozoic era so much biomass remained unburned that it was simply buried, to be dug up or pumped out later as coal and petroleum.

The characteristic pattern of burning on the land is called a *fire regime*. This concept qualifies in important ways the expression that such-and-such a species is "adapted" to fire, which is akin to saying that such-and-such a species is adapted to water. What matters is whether the organism can accommodate a pattern of burning, for if that pattern changes, the species may no longer be fire-adapted, any more than a species that thrives under consistent rainfall that comes month by month can survive if the rainfall is divided between distinctive wet and dry seasons.

Fire's patterns are complex—as complex as the ecological context within which it burns. A fire regime is thus a statistical composite. Within a regime, fires occur much as storms do within a given climate. Typically, there is not one kind of fire that happens only at one time of the year, but swarms of fires at various times.

Just as Phoenix and Cleveland may experience rain showers and thunderstorms, the way those storms are organized over seasons and years creates very distinctive climates for Arizona and Ohio. So it is with fire.

Until recently, there was one stunning lapse in life's control over fire. It could not govern ignition. That changed, however, during the Pleistocene with the arrival of hominids capable of manipulating fire and equipped with implements to spark it. It appears that *Homo erectus* could maintain fire once kindled, and that *Homo sapiens* could manufacture fire more or less at will. The arrival of a fire-wielding species was a monumental moment in the natural history of Earth: one—and only one—species acquired that capability. Its monopoly conveyed a peculiar and particularly disruptive power. In short order, humans found they could bring fire nearly everywhere, and they have done so.

The ecology of fire today bears little relationship to what preceded that evolutionary spark. For many observers, the human use of fire—the Promethean event—seems an insult to nature's (presumably) equilibrium fire regimes. But for others, the appearance of humanity as a fire creature is itself a logical, evolutionary event, one that completes the cycle of fire for the circle of life. Whether the outcome is "natural" is a cultural judgment, but it is certainly more biologically complete. Through people, life has strengthened immensely its grasp over fire on Earth.

CHAPTER ONE

The Fire after the Ice

What a purely natural state of fire might be in America is unknown, probably unknowable, and in any event hypothetical, since the present geography of the continent developed only after glacial ice had receded, ancient lakes had shrunk, and immigrant flora and fauna had begun colonizing the newly exposed lands. Yet these climatic upheavals occurred with fire-wielding peoples already on the scene. What matters (and what is truly interesting) is not how a purely natural regime might look, but how people and nature have interacted across so many millennia.

In some cases, they complement each other. Human-caused, or anthropogenic, fire may occur at the same times and in roughly the same places as lightning fire, simply adding to the overall load. In other cases—far more often—they compete. People burn outside the lightning season and in areas not regularly subject to lightning fire; they burn early, they burn over weeks or months, and they burn along routes of travel that may have little logic for lightning. These fires establish new conditions within which lightning fire must operate. People destroy the old patterns and impose new ones that show scant continuity with what went before. They may suppress lightning fire, convert forests to grasslands, or remake them into fields and villages, or suburbs, exurbs, and shopping malls. The fire regimes of America prior to European settlement displayed all these options— fire as complement, as competitor, as creator.

Still, two general patterns of human burning are apparent. One is characterized by control over ignition alone. People kindle fire on the landscape and, within limits, stop it. They may exercise some influence over that landscape and its fuels (say, by helping kill off megafauna), but they control fire primarily by controlling how, where, and when it starts. Call this state *aboriginal* fire practices. In the second pattern, people manipulate fuels as well as ignition. They cut plants and let them dry, or otherwise ready landscapes to receive their fires. They convert raw biomass into available fuel. These are practices commonly associated

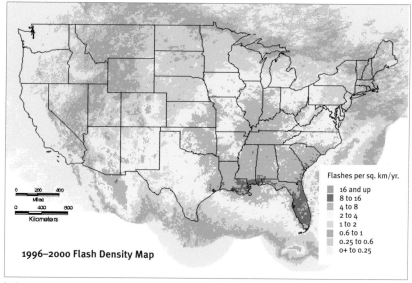

[1a]

Figure 1. Lightning-caused fires. Lightning flashes (a) do not always cause forest fires (b). The two phenomena overlap in a significant way mostly in Florida. Where lightning ignitions are most frequent, fires tend to be small, so the number of starts is not an indication of the acreage burned. But the second map can serve as a crude proxy for a natural geography of fire in the United States.

with farming, so call this suite of practices *agricultural fire*. All American tribes practiced aboriginal burning, and those that boasted agricultural fire also conducted aboriginal fire on lands not subject directly to cultivation.

ABORIGINAL FIRE

To master ignition is a mighty power. Unlike an axe or a spear, which can fell a single tree or bison, fire can spread and interact with many other processes. It can shape whole landscapes—a power far beyond the capacity of stone scrapers and arrowheads. (Think of it as a computer virus that can rewrite the operating software behind an ecosystem.) Yet for all its potency, aboriginal fire has formidable limitations. Its power derives from its ability to propagate, which means it can start and spread only where the landscape will support self-sustaining flame. A landscape that is too wet, snowy, or rocky will not carry fire, no matter how many sparks are cast into it. Most ancient American fire legends reflect this understanding. They tell how some mythical hero, having stolen the precious flame,

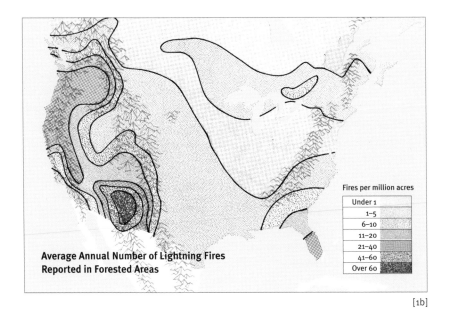

Fires per million acres

| Under 1 |
| 1–5 |
| 6–10 |
| 11–20 |
| 21–40 |
| 41–60 |
| Over 60 |

Average Annual Number of Lightning Fires Reported in Forested Areas

[1b]

distributes it about the landscape, whence fire must be rekindled by striking flint, or rubbing wood, or blowing on coals.

Aboriginal fire thus worked best in landscapes that already held fire or that had the essential ingredients for it. For the first, people could easily seize control over fire by substituting their ignitions for nature's. For the second, where the circumstances favored fire but lacked ignition, they could insert fire into the scene. But just as a change in landscapes can change fire regimes, so a change in fire practices can change landscapes by shifting the season for burning—from summer to spring, for example—or changing the frequency of burning or its scale; humans can reconstitute fire regimes and in this way control ecosystems. By their firepower, aboriginal societies in the right circumstances could shape their habitats. In fire-intolerant sites, fire stayed in the hearth. In fire-prone sites, it could remake whole landscapes.

To what ends did American Indians use fire? To any that could serve their purposes. They used fire as a technology—to harden spear points, to fire pottery, to chip stone, to light dwellings, to cook. They used it on the landscape in ways both tiny and huge. With torches they could clean ground to better collect acorns and chestnuts; they smoked bear and raccoons out of dens; they attacked enemy encampments; they cleared trails; they stimulated edible tubers (such as camas); they pruned and promoted berry patches; they fired dense woods to create accessible firewood. With torches they fished at night. By burning around wetlands, they encouraged habitat favorable to ducks and muskrats. By selective firing in sedge and shrubs, they promoted thatch and twigs suitable for baskets. With smoke they could attract deer and elk driven mad by flies. By burning, they kept lands around settlements and houses open, which prevented ambush by foes or predators and shielded them from wildfire. A land unburnt was a land uncared for—a land that was uninhabitable.

The most spectacular use, however, involved fire hunting. The ecology behind it involves both a push and a pull. By setting fires and letting them run with winds, hunters could drive animals into sites where they could be harvested; those burns, in turn, freshened browse and grass that, when greened, attracted more game animals and allowed more specialized hunting. Done properly, the fire hunt was indefinitely renewable. Its variants are many: almost any animal that *could* be hunted with fire as a means to repel and attract *was* so hunted; rabbits in California, woodrats along the Colorado River, bison on the prairies, deer nearly everywhere. In the Great Basin tribes practiced fire surrounds for grasshoppers. In Alaska they torched rings of spruce to drive moose. In northern wetlands they burned to expose muskrat trails, and in southern wetlands, to reveal alligators. Writing in the Carolinas in 1700, John Lawson noted that the natives "common go out in great Number, and oftentimes a great many Days Journey from home, beginning at the coming in of Winter" to fire-hunt and, that "they go and fire the Woods for many Miles, and drive the Deer and other Game into small Necks of Land and Isthmus's, where they kill and destroy what they please." Cabeza de Vaca described a similar pattern in 16th-century Texas. "Those from further inland…go about with a firebrand, setting fire to the plains and timber so as to drive off the mosquitoes, and also to get lizards and similar things which they eat, to come out of the soil. In the same manner they kill deer, encircling them with fires, and they do it also to deprive the animals of pasture, compelling them to go for food where the Indians want."

There were patterns to the practice. A simple way to imagine the outcome is to think of lines of fire and fields of fire—"lines" referring to routes of travel, and "fields," to those sites where burning was routinely practiced to promote foraging, hunting, and fishing. Thus, people burned along corridors of seasonal movement, they burned both deliberately and accidentally, and as they moved through the landscape, they burned the same patches, each in its proper season. Typically, sites in a condition to burn were burned: they were burned early, often, and lightly. But people did not always return with exact regularity, and fires did not always stay in place. In this way the burning varied, year to year. Besides, a good deal of fire "littering" occurred—fires left on the land in normal years burned themselves out, but in droughty years they could bolt across the scene. How these general patterns expressed themselves varied with the particular biota and setting in which they occurred. Common techniques applied in tall-grass

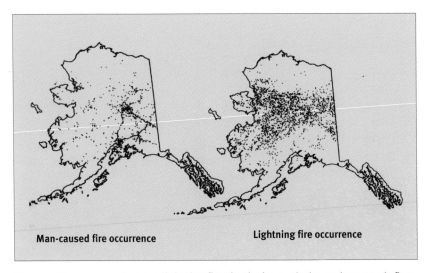

Man-caused fire occurrence   **Lightning fire occurrence**

Figure 2. Human-caused versus lightning fires in Alaska. In Alaska, anthropogenic fires track routes of travel and inhabited places (left), whereas lightning fires are largely restricted to the mostly uninhabited Yukon Valley (right). As people occupy territory, they change the environment within which lightning fire must operate.

prairie could result, for example, in very different outcomes when used in boreal forest.

Since human fire interacted with everything else in the landscape, as the landscape and its species changed, so did fire. Of particular interest is the disposition of animals that competed with fire for fine fuels, like grasses and shrubs. In the post-Pleistocene era, the extinction of large mammals led to two opposing outcomes. In fire-prone places, their disappearance allowed fuels to build up, thus furthering humanity's firepower. But in fire-intolerant places, the loss of animals that had kept a landscape open might make fire more difficult: it might cause the tree canopy to close in and shut down the conditions for fire. In America both trends happened until Europeans introduced domestic animals—horses, cattle, sheep, goats, and swine—that reversed the process.

The exact character of aboriginal burning at the time of European contact is difficult to recapture today. Those Europeans who were first on the scene had concerns other than fire practices and their effects, and by the time 18th-century Enlightenment naturalists arrived to inventory plants and animals, a century or two of profound change had altered the original conditions. The indigenous populations had declined and in many cases collapsed under the blows of slaving, war, forced migration, and especially disease; the landscapes they had cared for reverted to forest, and fire followed. Domesticated fire went feral.

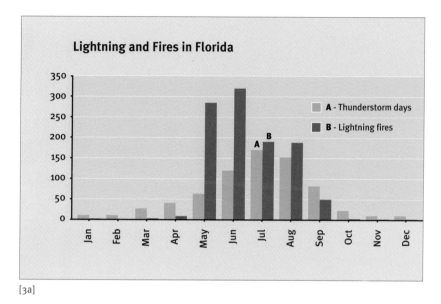

[3a]

Figure 3. Human-caused versus lightning fires in Florida. In the typical pattern, lightning fires (a) are most common at the start of the rainy season, with the right mixture of wet and dry. Those same fires can be graphed within an annual cycle of anthropogenic burning (b), which occurs mostly outside the season for lightning fire. The number of human-ignited fires dwarfs the number of fires ignited by lightning. The data come from protected forests in Florida, but the pattern holds around the world.

Better records are available for those places, like Australia, where European contact was delayed until the 18th century, and aboriginal fire and Enlightenment naturalist could meet along a flaming front. Such records testify to the power of aboriginal burning. As the Australian anthropologist Sylvia Hallam put it, the land at the time of European contact was "not as God made it but as the Aborigines made it." That was no less true in America. Explorers often marveled that the land was so often in a condition ripe for human occupancy, as though silently prepared for Europe's emigrants. That was so because humans had, in fact, already occupied it and through long centuries rendered it habitable, not least by their use of fire.

AGRICULTURAL FIRE

To increase their firepower, humans had to improve their control over the conditions that restricted aboriginal fire. This meant manipulating fuels and remaking the landscape so that it could better accommodate fire. Globally, this happened

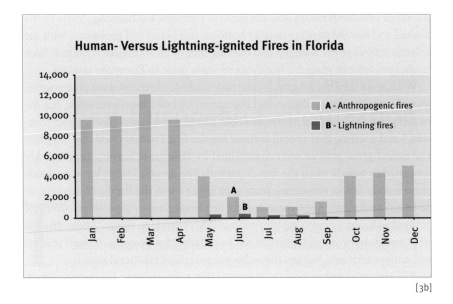

Human- Versus Lightning-ignited Fires in Florida

A - Anthropogenic fires

B - Lightning fires

[3b]

through such measures as cutting and drying plants, draining wetlands, and unleashing livestock, all of which converted incombustible biomass into stuff that could ignite. In pre-Columbian America, where domestic livestock were absent and large-scale draining was not done, this meant slashing or killing woody plants and, after letting them dry, burning them to ash. This practice expanded both the geography of fire and its seasons. People could now apply fire in places and at times that would not otherwise burn; these effects varied as human populations rose and fell.

Agricultural fire was an exercise in applied fire ecology. In wildland fires the greatest effects occur in the first year after a burn; by the third year, most of the recovery is well underway. This describes exactly the rhythm of slash-and-burn cultivation. You can plant in the ash, often with spectacular results. But a second crop a year later is tricky, and by the third year the native plants and animals have reestablished their grip, and introduced crops will struggle. By then, however, a new site will have been readied and fired. Over the years the original site will be revisited; this time it will be much easier to clear and burn. In this way patches and pulses that describe fire's "natural" ecology are disciplined into the fields and rotations of farming.

Variations of such a system, based on Mesoamerican crops such as maize, beans, and squash, spread throughout southeastern America and northward in the Ohio Valley, the wetter drainages of the Mississippi, and the Northeast shy of the boreal forest. Excavations in the Tennessee Valley have tracked the patchy

in the Mediterranean. For all that, the Spanish little colonized what became America.

That task fell to emigrants from temperate and boreal Europe, which by the 18th century had become the main source for imperial colonization, science, and industry. The inhabitants of France, Germany, Britain, and the Netherlands all thrived amid a climate that had no routine dry spells. Temperature, not precipitation, changed through the seasons, so natural fire did not exist, and although burning was integral to all varieties of European agriculture, it occurred primarily because people put it there. Even boreal Europe sought to make its lands conform to a temperate model. The landscape as garden was everywhere the European ideal and norm.

In fact, temperate European agriculture was a fire-fallow system: fields rotated through a regular cycle of cultivation during which they were at some point left fallow, and then burned. Agronomists hated fallow, which they deemed wasteful, and since it involved burning, they condemned fire as well. Peasants burned, so agronomists reasoned, out of superstition and tradition, whereas a "rational" agriculture would find substitutes for fire and thus put those fallow "wastelands" into cultivation. The peasant farmer had a more reasonable explanation: agriculture demanded burning, just as initial land clearing did, and fallowing was necessary to ensure a sufficient amount of fuels for that catalytic fire. The hostility of agricultural ministers and academic agronomists toward fire matters, however, because forestry evolved at the same time, as a graft on the great rootstock of European agronomy, and it regarded fire in the same hostile way. Foresters hated and feared fire and joined other elites who denounced it with a mixture of despair and outrage.

What emerged out of contact between Old World and New was a hybrid. Europeans used fire to clear and to promote arable fields and rough pastures, much as they had in Europe. But those folk on the frontier quickly adapted fire practices from the Native peoples as well. The earliest fire codes, for example, sought to limit fire hunting, which seemed both destructive and unsporting. Many farms grew on abandoned fields originally fire-cleared by the indigenes; many pastures reclaimed old hunting grounds. Of special interest is the hybrid fire economy that emerged from the Swedish colony on the Delaware, where Finnish slash-and-burn farming and traditions of long hunting merged with Lenape equivalents and spawned a defining backwoods frontier society that spilled over the Appalachians and across much of middle America.

There were differences, of course. Wildlife did not overgraze, but livestock did, and so burning assumed somewhat different patterns. Land clearing for logging and farming could slash new forests and leave colossal mounds of debris behind, occasionally sparking monstrous fires. And landownership favored farming on fixed sites. The old fire regimes had resulted from peoples who moved through varied landscapes seasonally, burning each according to need and availability. The new regimes demanded that people stay put. Instead of farms circulating through the land, the land (as it were) rotated through the farm in a succession of planting, fallowing, and burning. Only where open commons on a large scale endured did pastoral burning and something like aboriginal fire survive. As it rippled across the New World, the process of replacing one set of fire practices with another was rapid and reckless, and to the thoughtful, extravagant and embarrassing.

The actual mix and details of that exchange varied. For the western United States, the process involved the removal of Native populations followed by massive overgrazing. The stocking by sheep and later cattle stripped the grassy cover that had sustained the existing fire regimes; a woody scrub substituted. In the Southeast, one mixed fire economy succeeded another but with grazing added, along with an often-destructive passion for plantation cropping that exhausted soils and encouraged erosion. For most of the eastern United States, Europeans undertook landscape-scale clearings and drained wetlands, both of which exposed vast stocks of new combustibles to fire. These burned, frequently in the autumn, giving rise to New England's famous Dark Days (May 19, 1780, was perhaps the most memorable) and, worse, to great fires that blazed across immense tracts of land. The defining episode was the Miramichi burn of 1825, located mostly in New Brunswick, but with fraternal fires in Maine and coastal New England, which collectively burned nearly four million acres. That pattern continued for another century, with each new frontier of axe and hoof breaking out in extensive burns while the nation's rapid industrialization quickened the tempo of change.

This disruption of fire regimes put fire into landscapes where it had not been significant before and removed fire from landscapes where it had been common. Most spectacularly, the frontier became itself a flaming front that occasionally exploded into large destructive fires. Such eruptions further convinced elites, already suspicious of burning, that fire was intrinsically harmful and occurred because of human malfeasance. America's first professional forester, Bernhard Fernow, famously attributed America's fire scene to "bad habits and loose morals." That fire might serve a legitimate role in cultural landscapes and a necessary one in quasi-natural settings seemed irrelevant when abusive burning spilled across landscape after landscape. Critics linked almost every environmental ill and insult to fire as a cause, consequence, or catalyst.

What did the geography of frontier America look like? Helpfully, three samplings exist. In 1878 the Geographical and Geological Survey of the Rocky Mountain Region under the direction of John Wesley Powell produced a map

[4a]

Figure 4. Geography of fire in late-19th-century America. In 1878, John Wesley Powell directed a survey of the Rocky Mountain region and produced a map of Utah (a) that indicated extensive burned area (shown in tan). In 1880, Charles Sprague Sargent, director of Harvard's Arnold Arboretum, produced a map of fires (b) as part of his survey for the 1880 U.S. Census report on forests. The darkest color indicates the highest percentage burned annually (more than 10 percent). Some areas were left blank because there were no forests or data were insufficient.

of Utah that included burned area. In 1880 Charles Sargent, a botanist and director of Harvard's Arnold Arboretum, drew a map of forest fires for the entire country as part of his survey of forests for the census. And in 1882 Franklin Hough, appointed the country's "forestry agent" in 1876, published his massive compendium on the nation's forest estate, recording a dense digest of its fires. Together they yield an enlightening cross-section into the fire scene at a time when industrialization and settlement were together shredding the old cartography of burning.

The 1878 *Report on Lands of the Arid Region of the United States* has long enjoyed status as a classic of American conservation. Its larger purpose was to help rationalize western settlement by exposing the full meaning of aridity as an implacable feature of America West of the 100th meridian (along which annual rainfall approximates 20 inches). The West's periods of annual dryness and longer-

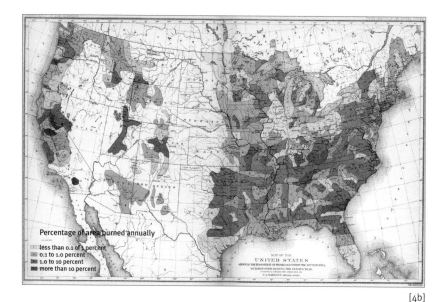

[4b]

term drought made fires possible in ways not characteristic of the more uniformly humid East. Yet climate was only half the equation: there had to be ignition as well. Powell's crews would have found lightning significant except that its contributions to fire regimes were overwhelmed by the presence of human burning. Powell concluded that "in the main these fires are set by Indians," primarily "for the purpose of driving the game." Such facts, he added, were well known to those on the scene. By now Powell had himself become an authority on the Piutes.

The magnitude of burning was staggering—"on a scale so vast that the amount taken from the lands for industrial purposes sinks by comparison into insignificance." At risk, however, was more than timber, since those forested mountains were the watersheds that would make possible the irrigation of the arid valleys. The forests had to be protected, and protection meant little unless it included fire control. Powell, who then headed the Bureau of American Ethnology, concluded that the burning could be "very greatly curtailed by the removal of the Indians." And so it proved.

Charles Sargent's map had less to say about the West and almost nothing about the High Plains, since its theme was forests, which in the West were scattered and in the Midwest replaced by grasses. Unlike Powell's fire map, which showed only whether a landscape was burned, Sargent's showed proportions of land burned annually. Even a casual glance shows little correlation between where lightning was predominant and where the burning occurred. There is overlap only in the greater Southwest, parts of the interior Northwest, and Florida, but this reflected favorable conditions for burning, not simply ignition. Burning corresponds to places where humans remain important fire agents, and the geography of fire in 1880s America reflected its use for land clearing, fire-fallow farming, and stimulating grasses for grazing sheep and cattle: the United States of the day was an agricultural society. That it was also rapidly industrializing helped stir up yet more opportunities for fire and fuel to meet.

One striking fact is the persistent frequency of fires in the Southeast. Almost certainly this was true prior to European contact, since both agricultural and aboriginal fire flourished; it continued in the 1880s because the rural economy required fire and climate favored it almost year-round—and fire remains frequent there today, in the form of controlled burning. In contrast, the situation in the Northeast and the Lakes States has changed. In the 1880s, both regions witnessed huge fires feasting on logging slash; a century later, fire had almost completely vanished from both regions, save in patches of reserved boreal forest.

Franklin Hough's *Report upon Forestry* adds further details. It documents a use of fire that was both pervasive and casual among the general population. Fire was what made their world habitable, from the hearth to the field burn. It was what shielded them from wildfire, be it from nature, accident, or arson. Those folk living on the ground exploited fire wherever and whenever it could help. And if it occasionally escaped because of drought, wind, or logging debris, that was a small price to pay for progress. Those wild outbreaks would cease when the frontier ripened into farms, villages, and meadows.

These texts and maps could thus be read differently by those living on the land and those who only studied it. As in Europe, intellectuals and officials found fire alarming, and its dangers far more impressive than its benefits. The folk saw fire as essential and generally benevolent, and its hypothetical removal both delusional and destructive. They wanted fire and were willing to tolerate occasional breakdowns to keep it. The elites distrusted fire and saw those fatal conflagrations and incinerated forests as a cost too great for a civilized country to bear. They wanted something they came to call conservation, and conservation, they understood, could work only if it controlled fire. The free-burning ways of the frontier had to end.

CHAPTER THREE

Fire and Conservation

I n 2006 the U.S. Geological Survey published a map of large fires from 1980 to 2005. The map bears as little relationship to Sargent's 1880 map as Sargent's map does to a map of lightning fires. The large fires (defined as a burn of more than 100 acres) do not include controlled burns, which continue to flourish in the Southeast. But the map does reveal the geography of the contemporary U.S. fire scene and shows how, over the course of a century, the history of American fire has virtually inverted its geography. Regions of massive fire such New England and the Lakes States no longer have them, and western areas of modest burns have acquired large ones. How has this happened?

The short answer is industrialization, the creation of a public domain, and the invention of institutions to oversee fire. Each of these factors has contributed to altering the geography of fire, each has its own internal history, and each displays its own dynamics. The industrial narrative is what America shares with the rest of the world. The public domain narrative is something it shares with nations such as Australia and Canada. The institutional narrative is unique to each country. Together the three narratives braid into a master pattern. The distinctive way they have braided describes fire's history, and their resulting pattern, its geography.

INDUSTRIALIZATION

Industrial combustion is today the primary driver of fire on Earth. Much as agricultural fire overcame the primary limitation on aboriginal fire (available fuel), so industrial fire has overcome the principal check on agricultural fire, its self-limiting reliance on surface biomass for adequate quantities of stuff to burn. Even with chopping, draining, planting, and so on, only so much potentially combustible material can be coaxed from the land; if overdone, the biota cannot be replenished fast enough—and there are plenty of examples of such

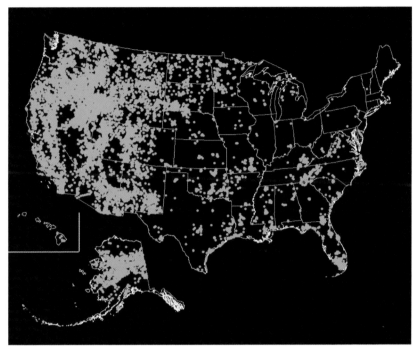

[5a]

Figure 5. Contemporary geography of large fires. Except in parts of the Great Plains, the distribution of large fires (a), as recorded by the U.S. Geological Survey, 1980–2005, is very largely a map of public lands. The distribution of prescribed fires (b) is largely concentrated in the Southeast.

[5b]

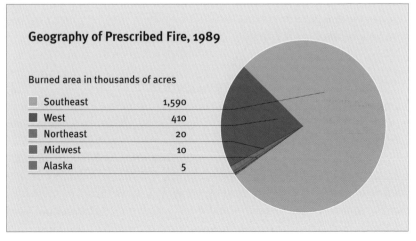

Geography of Prescribed Fire, 1989

Burned area in thousands of acres

▨	Southeast	1,590
■	West	410
■	Northeast	20
■	Midwest	10
■	Alaska	5

degradation. To increase their firepower, then, people had to discover another source of fuel, which they did in the fossil biomass of oil and coal. They could burn landscapes from the geologic past. In this new fire frontier, their firepower became virtually unbounded. The combustion of fossil biomass is, for fire history, the definition of *industrial fire*.

Industrial fire competes with open burning in several ways. One means is by technological substitution. To the extent that people consider fire a tool, they can get the same effects without open flame and its unwanted side effects, like smoke, escaped burns, and the inefficient consumption of woody fuel. They can use machines powered by internal combustion to heat, cook, weld, smelt, light rooms, mow lawns, and clean yards. Even agriculture has turned away from fire as a source of fumigation and fertilization in favor of chemical pesticides, herbicides, and fertilizers (in effect, using fossil biomass as a kind of fossil fallow). In setting after setting, from homes to cities to factories to fields, open flame has steadily disappeared as alternative tools have been devised. Leaf blowers replace burning piles and wood chippers substitute for slash burns, much as electric bulbs replace candles and gas heaters the family hearth.

The process of conversion—what we might term the "pyric transition"—may be violent and chaotic, but it matters because it provides the backdrop for much conservation thinking regarding fire. In rough terms, what happens is that the population of fires behaves much like the population of people. As industrialization ramps up, a period of explosive growth begins. The old ignitions endure, and new ones are added, all stirred into landscapes upset by the access made available by new modes of transportation like railroads, and in the case of still-colonizing countries, the general havoc of settlement. The population of fires multiplies, and their damages swell. Eventually, technological substitution and active suppression (made possible by new fire machinery) cause the numbers to plummet and may drive them below replacement values, which for fire means there is not enough burning to do the ecological work required.

During the Great Barbecue, as V. L. Parrington called America's post–Civil War era of cronyism, resource exploitation, and all-out industrialization, the nation underwent just such a violent transition, with massive fires consuming forests and from time to time overrunning settlements. Yet paradoxically, the prevailing sense was that the fires would go away by themselves, that the very act of settling land would remove the conditions that fed fire. The tragedy of pioneering—that pioneers destroy the circumstances that make pioneering possible—would by itself abolish wildfire. Wild fires would go the way of wild lands. But many observers saw no more reason to believe that untrammeled settlement would achieve its goals than that unshackled capitalism would bring wealth and justice to all. They demanded that government intervene for the sake

[6a]

Figure 6. The Great Barbecue. The 1871 Peshtigo fire (a) was part of an immense regional complex that included the burning of Chicago. Here an artist depicts refugees seeking protection in water near White Rock, Michigan. The railroads (b) cracked open the North Woods beginning in the 1870s. By leaving slash and strewing cinders, they created the perfect conditions for large-scale fires, like this one in Minnesota in 1923.

[6b]

[7a]

Figure 7. The Great Barbecue, continued. The era of big fires began with agricultural settlement and did not cease until the 1930s, after the great droughts faded and government programs to alleviate the Depression beat down fires. The Yacolt fire of 1902 (a) was Washington State's largest; 20 years later, when this photo was taken, the wreckage was still visible. Virginia's Great Dismal Swamp (b) was drained by drought and burning; the Virginia Department of Forestry in part exhausted its fund by 1931 by trying to halt an earlier fire in the area that burned from 1923 to 1926 and consumed 100,000 acres.

[7b]

of the commonwealth. Their solution was to apply the technologies of industrial combustion, along with the power of the state, to deliberately suppress its wild rival.

EARLY MODELS FOR FIRE PROTECTION

When Americans consider other countries that seem to have similar fire problems, they look to Australia, Canada, and maybe Russia. All have extensive fires, fire management agencies, and fire research institutions. Although each has a distinctive national character, they share related concerns about wildland fire for a common reason: they all have extensive wildlands. This fact reflects not simply common fire climates and fire-prone biotas but a common history of colonization that bonded with a philosophy of state-sponsored conservation. Specifically, during their settlement surge, they all set aside large estates from which the indigenous peoples were removed, either deliberately or by disease. Those newly vacant lands remained under the control of the national state.

The precedents for reserving vast tracts of land lay in European colonization overall, and Americans looked to European imperial powers, notably Britain and France, for inspiration. The similarities among today's fire powers is the result of a deliberate, global project of creating forest reserves among colonized lands and of the shared values of a transnational corps of foresters. The idea of reserving lands was an axiom of global conservation: it was the most direct means one could take to end the wreckage that settlement set into motion. The usual charge was to protect the forests "from fire and axe." The critical distinction among lands set aside, however, was between those that were uninhabited and those that remained within the daily use of resident or adjacent populations. Where lands remained uninhabited, forest reserves more or less succeeded as intended. Where people retained access to forage, graze, farm, or hunt, they generally did not. Those extensive wildlands that made possible America's extensive wildland fires are thus a historical anomaly.

By the late 19th century the United States had three broad categories of public land reserved for conservation purposes. There were reserves created by individual states, notably the Adirondacks and Catskill preserves in New York. There were national parks, beginning with Yellowstone in 1872 and expanding into the Sierra Nevada during the 1890s. And after 1891, there were national forest reserves that could be carved out of existing public lands, by presidential proclamation. In time there were other categories of public land, most importantly national monuments, created by presidential proclamation from the existing public domain under the 1906 Antiquities Act, but also wildlife refuges, and after the Weeks Act of 1911, national forests acquired by purchase. But the

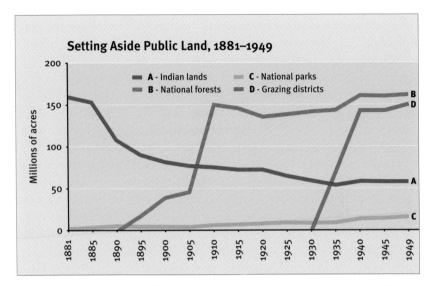

Figure 8. Public domains, public fires. The creation of public lands, beginning in the late 19th century, required the federal government to become an active steward; in the early years, this meant protection from timber trespass and fire. Today, about a third of the national estate is public, with a third of that in Alaska.

three initial categories were the ones that matter most because they provided the models for fire protection.

Still, there were other options—timber protective associations and rural fire brigades, for example. The timber protective associations arose most prominently in Idaho (and were known for years as the "Idaho idea"). In part, they reflected traditions of collective action on the frontier by which neighbors banded together for mutual aid. But the state helped kickstart the programs by threatening (and in many cases enacting) legislation that required landowners to provide fire protection or else face state taxes to fund an agency to do it for them. The voluntary associations flourished best where federal lands were weakest; they took root especially in New England (and eventually formed the primary basis for fire protection in Quebec). Still, even associations could not reach beyond their members. Only national efforts could command a national system.

INSTITUTIONS AND IDEAS

Initially, it seemed enough to shield public lands by legally setting them aside as reserves, which removed them from the frontier of settlement. But actually stopping poaching, grazing trespass, and fire demanded a more active administrative presence: it required policies, institutions, actions on the ground. From each of the major categories of reserved lands arose a different model.

The Adirondacks version evolved from rural fire protection and relied on an existing infrastructure: wardens could call upon local crews, supplemented by outside help through a state forestry bureau. The major fires that ripped through the preserve in 1903 and 1908 dominated regional news, which is to say, national news as well, and the New York experience provided inspiration for other states throughout the Northeast and around the Great Lakes.

The national parks required something else. They could not rely on local communities, and they suffered from an administrative paradox in their charters, which identified their purposes as equally to protect nature and to promote access for the public. Whereas most other countries brought new national parks into a system through the common provisions of a single enabling act, the United States created parks through individual acts of Congress, and there was neither a collective identity for the lands nor an agency to administer them. In 1886, frustrated by the incapacities of an inadequate civilian overseer, Congress assigned administration of Yellowstone Park to the U.S. Cavalry. The troops remained for 30 years, until a 1916 National Park Service Act established a civilian agency to attend to the growing collection (it was too amorphous to be called a system). The other major parks also fell under Army administration. When M Troop rode into Yellowstone in August 1886, they were greeted with flames; they extinguished 60 fires altogether that summer, establishing a paramilitary model of federal fire suppression that continues in muted form today. Fire was indeed something to fight, to contain as one would a hostile force.

Both state and national parks, however, were small, and their influence mostly symbolic. The critical contest involved the forested lands on the public domain, especially those designated as forest reserves. The task of overseeing them originally fell to the General Land Office, whose primary duty was to dispose of the public domain—hardly an agency prepared to actively manage lands that the government wanted to retain in perpetuity. But again the enabling legislation (a rider to the 1891 Sundry Civil Appropriations Bill) allowed the president only to declare reserves; it said nothing about governing them or paying for their administration, much less designating an agency to do so. As the reserves expanded, the search for a suitable mechanism heightened. The task was further complicated by uncertainty about just what the lands were to do other than be "protected."

In 1896 the National Academy of Sciences was called upon to staff an inquiry. The Committee on Forests' chairman was Charles Sargent, and among its other members were Arnold Hague of the U.S. Geological Survey and Gifford Pinchot; John Muir participated in an unofficial capacity. The committee toured select reserves in the Northwest, and for six weeks through six western states it was never out of sight of smoke. Among prospective models for management, it investigated New York's approach and the military in the national parks as well as British India. In the end, the committee proposed that for the near future, management of the reserves be turned over to the Army, just as the parks had been handed over to the U.S. Cavalry. To assist, it recommended that graduates of West Point be trained for service in a corps of forestry engineers. It further urged a scientific survey of the reserves under the auspices of the U.S. Geological Survey.

In the end, Congress enacted parts of the proposals as the Organic Act of 1897 and authorized the U.S. Geological Survey to map coarse forest and burn features, much as Powell had done in Utah. This time, the cartographers identified many sites repeatedly burned, a condition that all parties recognized as far more insidious than a single burn, however far-ranging. The reconnaissance amplified the impression the Committee on Forests had received on its summer fly-by: the woods were in bad shape, even those spared the axe, because of fire. John Muir voiced collective wisdom when he calculated that fires ruined ten times as much forest as logging.

The fires were everywhere, and although they differed—some savage in their fury, others little more than spring freshets that swept off the annual growth of surface debris—they were, so critics lamented, all regrettable and nearly all preventable. Most stemmed from human practices, which could be halted by education and the enforcement of proper codes. Some would still happen, if only by accident, and some would be kindled by lightning; these could be extinguished by quick response. In this sense, the nation's forests were little different than its cities. Good laws, good policing, and good firefighting could end the dark spiral of destruction.

Still, that ambition was not exactly policy, and on policy, even wise men might disagree. The widespread enthusiasm that characterized most folk burning found a surprising champion in John Wesley Powell. In 1890 Bernhard Fernow and Gifford Pinchot, the nation's two professional foresters, arranged to meet with Secretary of the Interior John Noble to present the case for forest protection. The best cause was of course fire protection—and it was the easiest argument to sell to the educated elite.

Powell crashed the scene, however, and according to a sour Pinchot, launched into a justification for a kind of Indian burning he had witnessed in his study of the western tribes. Regular surface firing did not destroy many western forests, he insisted. Rather, it kept the understory dampened so that it rendered such forests virtually immune to crown fires. This flew in the face of forestry doctrine, which had behind it academic standing as well as European experience and colonial experiments. By challenging the assumption that fire, all fire, had to be suppressed as a precondition to scientific management, Powell had called into question the legitimacy of forestry as a source of expertise and foresters as the proper guardians of any system of forest reserves.

Pinchot outlived Powell, although both men fell afoul of their bosses and were eventually driven from office. Likewise, foresters steadily advanced the claim that they, and they alone, were competent to administer the wooded public domain. They were the engineering corps for timber and watershed, just as other engineers were experts in mining, surveying, reclamation, and bridges. Their perspectives eventually triumphed, along with their values and notions of how forests should work and their conceptions of suitable fire regimes. But at the moment of creation, as the reserves were being conceived as a national project, this was not obvious. Foresters had no special warrant, and they met any challenge to their authority with savage denunciation. In particular, they remembered Powell's description of Indian burning, which they dismissed contemptuously as beneath the dignity of a modern, scientific civilization. It was, they sneered, "Piute forestry." No serious program of conservation could build on such philosophical sand. But without estates of their own to administer, foresters had no political clout. They could only advise and clamor for public attention.

Still, despite the recommendations of the National Academy of Sciences committee, Congress declined to authorize the Army to administer the new reserves. The reasons were several: the impending Spanish-American War, clear lack of interest by the Army, a keen sense of mission among young American foresters. Less obvious is why Congress did not assign the task to the U.S. Geological Survey; perhaps Powell's preference for federal control of irrigation projects, which had displeased western members and their development allies, doomed any prospects of giving the federal government's premier scientific agency responsibility for land management. Instead, the General Land Office staffed a small corps of rangers while the government's foresters remained in the Bureau of Forestry, a purely advisory agency in the Department of Agriculture. This division endured until 1905, when President Theodore Roosevelt transferred the forest reserves to the foresters. The reserves became national forests and the bureau, the U.S. Forest Service. The modern era of fire administration dates from that moment.

CHAPTER FOUR

The Great Firefight

There was never a question but that whoever oversaw the national forests would fight fire. But the guild of foresters brought special zeal to the task. Like forest reserves, fire protection went global, and with cause: there was no point in establishing reserves if they simply burned down. The fraternity of foresters helped carry ideas from one place to another, such that similar concepts and practices characterized India, Cyprus, Australia, and the American West. The conviction prevailed that fire protection was mandatory, and that a "system," a protocol informed by scientific forestry and backed by the muscle of the state, would make it possible. Towering smoke columns, more-over, were a visible test of those presumptions. Even while still with the Bureau of Forestry, Gifford Pinchot had thundered, with abolitionist zeal, that "like the question of slavery," fire protection might be deferred "at enormous cost in the end," but "sooner or later it had to be faced." His successor, Henry Graves, chief of the Forest Service from 1910 to 1920, thought fire protection was 90 percent of American forestry. William Greeley, chief from 1920 to 1928, downgraded it to a mere 75 percent. From the beginning, then, fire protection was a political project that obsessed the agency.

Still, America's foresters hardly appreciated the full magnitude of the task before them. They assumed that the principal issue was to regulate human behav-ior, that controlling people—restricting what they could do on national forests and how they might do it—would throttle the fire problem. Quell slash, muzzle grazing, restrict campfires, eliminate broadcast burning by prospectors: such measures would knock the fire crisis down to manageable proportions. What remained, the accidental fire and the lightning strike, could be contained by a suitable infrastructure and system of patrols.

They appreciated, too, that fire had unrivaled value for public relations. Its apparent damages could be used to remind the public about the need for conservation: the threat from fire justified state-sponsored conservation, and fire-fighting was a dramatic gesture that bespoke active administration. In the early days most Americans identified the forest ranger as a frontiersman-turned-firefighter. In some ways they still do.

Chief Gifford Pinchot stated the matter succinctly in the 1905 *Use Book*, a pocket manual that laid out the regulations for public use of the reserves and described the duties of his rangers: "probably the greatest single benefit derived by the community and nation from forest reserves is insurance against destruction of property, timber resources, and water supply by fire." The "burden of adequate protection" was too great for citizens or states; "only the Government can do it, and, since the law does not provide effective protection for the public domain, only in forest reserves can the Government give the help so urgently needed." It was not simply that reserves required fire control, but that the perceived need for fire control required reserves. In 1909 the Forest Service told the National Conservation Commission that even though its forces were much too small, it offered "the best example in the United States of an efficient system of fire prevention and control." Fire protection was both necessary and possible.

THE GREAT FIRES

The next year, 1910, shattered that presumption. That summer wildfires swept the West but struck with stunning ferocity in the Northern Rockies, where some 3.25 million acres burned, most of it in one savage rush that became known as the Big Blowup. As the fires grew, so did the effort to fight them; by late August some 9,000 firefighters were on the payroll, along with most of the standing Army in the Northwest. On August 20–21 the Big Blowup roared through them. Some 78 firefighters died in six separate incidents, and the entire apparatus of fire protection collapsed.

The fires traumatized the young agency, which in January had seen Pinchot fired for insubordination and was still reeling from the political threat implicit in his dismissal. The agency went into debt for nearly a million dollars and would have expired in bankruptcy had not Congress upheld a 1908 law that allowed it to overspend its budget during fire emergencies, after which Congress would pass supplemental appropriations. The burns were followed by a suspension of rules governing logging that allowed for emergency salvage harvests and replanting. The callout, the firefight, the memorials, the emergency expenditures, the hurried salvage and rehabilitation—the Big Blowup thus encompassed the whole apparatus of 20th-century American firefighting.

More importantly, it became spliced into the genetic memory of the institution. Not only were the Great Fires Henry Graves's first major challenge as

[9a]

Figure 9. The Great Fires of 1910. The military style of 20th-century fire administration originated in the Great Fires. Ranger Ed Pulaski and most of his terrified crew of 45 survived the Big Blowup in August 1910 by sheltering in the abandoned War Eagle Mine shaft (a) on Placer Creek in Coeur d'Alene National Forest, Idaho. That same month, proponents of frequent light burning (conducted to prevent large fires) challenged Forest Service fire policy. Henry S. Graves, who had recently become chief of the U.S. Forest Service, was no stranger to the controversy. In 1896, while on reconnaissance of the forest reserves in the Black Hills as part of the Geological Survey, Graves took this photograph of a surface fire (b) in a ponderosa pine forest in Rock Pine, near Hill City, South Dakota, an area that became part of the Harney National Forest and then the Black Hills National Forest. The devastation of the Great Fires inspired a national enthusiasm that made firefighting akin to waging war (c). This photo, taken in July 1927 on the Lassen National Forest, California, shows firefighters going to the front at the Mineral Fire.

Pinchot's successor, but the next three chiefs were all personally on the scene with them. The fires became for the agency a defining moment in its character. Not until this entire generation passed from the scene in 1939 did the Forest Service begin to banish the specter of the Big Blowup and allow some small space for fire as anything but an enemy to be fought.

[9b]

The practical challenge came with a conceptual one as well. Even as the agency was hurling crews at the flames, an article appeared in the August 1910 issue of *Sunset* magazine that argued for an alternative strategy of forest protection. In essence, it revived Powell's appeal for the "Indian fires" that relied on frequent light burning to prevent large fires. This was also the means preferred by those living on the land—its ranchers, its timber owners, its farmers, the whole folk tradition that looked to fire to shape a more habitable environment. It argued for fire *lighting* rather than fire *fighting* as the basis for forest protection. Regular firing would prevent the need for mass callouts such as those in the Northern Rockies; conversely, if such practices were removed, the forests, particularly in the West, would succumb to insects, fungi, and catastrophic conflagrations.

Foresters of course recognized the political threat, for the issue was fundamental to the entire scheme of global forest reserves. The first question asked at the first forest conference in British India (in the 1870s) was whether fire control was possible, and if possible, desirable. If the answer to both questions was no, then the founding premises of state-sponsored forestry were flawed, and those administering the reserves would have to devise some other basis for their oversight. The debate turned, as in Europe, on the split between those in offices and those in the field. With few exceptions, officials demanded fire control while field

[9c]

operatives warned that this was not possible and that even the attempt might cause unintended consequences. For colonial officials there could be only one politically acceptable answer: suppress.

American foresters echoed the divide between office and field but with less hesitation from its rangers, for the founding generation was astonishingly uniform in age, background, and beliefs. Collectively, they fought the light burning debate, as it was termed, as vigorously as any backcountry smoke. When light burners found a champion in Richard Ballinger, secretary of the Interior and rival to Pinchot (a competition that had led to Pinchot's dismissal), the quarrel over fire policy became a proxy for deeper politics.

The Forest Service responded in two ways. One, it promoted "systematic fire protection" as its basic strategy. In 1914 Coert duBois published a model system for the California forests that gradually became the national norm. And two, it denounced light burning with all the authority at its command. Henry Graves and William Greeley both derided the proposal as "Piute forestry." Writing from the Southwest in 1920, Aldo Leopold concluded that "the Forest Service policy of absolutely preventing forest fires insofar as humanly possible is directly threatened by the light-burning propaganda," that repeated light burning had destroyed the forests' reproductive capacity, and that the ruinous effects of "Piute forestry" were visible throughout the Southwest and California. In 1923 a special forestry panel investigated the whole California controversy and condemned light burning as nonsense and heresy.

Timing also mattered. The Great Fires occurred amid an age of reform, a period of political activism and conservation enthusiasms, an era in which pragmatism promised to free practitioners from doubt and allow them to act in a world about which they had incomplete knowledge. Theodore Roosevelt championed a "life of strenuous endeavor"; William James, a founder of philosophical Pragmatism, publishing his final essay as the smoke from the Big Blowup changed the New England sun to a coppery hue, advocated a "moral equivalent of war." Alarmed at the militarism he saw in Western civilization, James urged a redirection of that martial ardor to constructive pursuits; just as there was a mechanical equivalent of heat, so there might be a moral equivalent to war. He believed, in particular, that fighting the forces of nature, a common enemy, could replace fights against other peoples.

The drama of the Big Blowup and the firefight against it captured exactly this sentiment. The image of Ed Pulaski holding his crew of panicked firefighters in a mineshaft—at gunpoint even—while the firestorm passed overhead merged ideal with deed. The Great Firefight spoke to folk as well as elites. In an age that hungered to act, firefighting tapped an enthusiasm almost moral in its fervor. Before it entered World War I, America went to war against fire.

Two other reforms completed the extraordinary aftermath of the Great Fires. In 1911, after years of haggling and shocked by the western fires, Congress enacted the Weeks Act. This allowed for the expansion of the national forest system through purchase; the lands of interest lay in the mountains of New England and the southern Appalachians, both critical watersheds. But it also allowed the states to organize among themselves, and it established a program of federal and state cooperation. Not surprisingly, the focus was fire protection. As states enrolled in the program, the federal agenda, which is to say the fire ambitions of the U.S. Forest Service, found a way to nationalize. The program underwent a significant expansion with passage of the Clarke-McNary Act in 1924. A parallel program of forest research commenced in 1916 and received congressional authorization in 1928 with the McSweeney-McNary Act. These moves left the Forest Service as virtually the only sponsor of scientific research regarding fire.

The firefight, the controversy over policy, and the enabling legislation to build institutions—these became the forge, the hammer, and the anvil of a national system of fire protection. Out of the ashes of the Great Fires emerged, almost fully realized, the American way of fire protection. The Forest Service oversaw it all and was well on its way to dominating every aspect of wildland fire management.

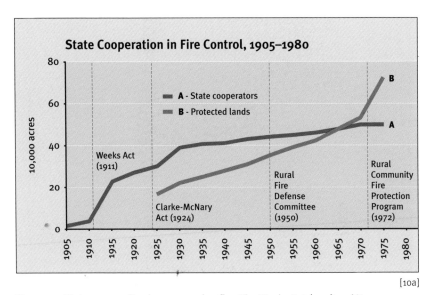

[10a]

Figure 10. State cooperation in suppressing fire. The Weeks Act (1911) and its successor, the Clarke-McNary Act (1924), brought the states into a national program of fire protection (a). As states signed on, the extent of protected land rose. Southern Ohio illustrates the immediate impact (b): the estimated fire load collapsed after organized fire suppression began in Ohio in 1923. Data in the graph prior to 1923 are estimates based on a forest survey in 10 counties. Fire statistics from 1943 to 2001 are from Ohio Division of Forestry records. Although fire protection programs are cooperative, the relative financial outlays by private, state, and federal institutions (c) indicate that the states have done the heavy lifting in removing fire from the landscape.

[10b]

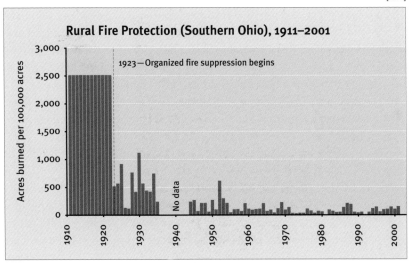

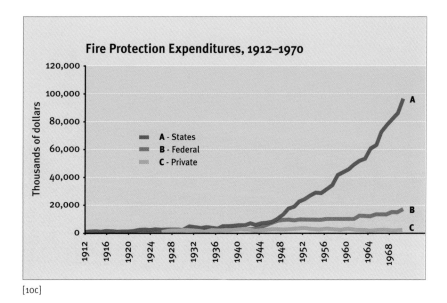

Fire Protection Expenditures, 1912–1970

[10c]

THEMATIC CYCLES

For the next 60 years the story of American fire is largely the unfolding of these themes. Fire was to be prevented where possible and fought where necessary. The means available to fire protection—its monies, its equipment, its crews—expanded, especially during the New Deal in the 1930s and after the Korean War in the mid-1950s. Despite occasional challenges, the inherited understanding of what fire meant endured.

Establishing a government presence on the public domain was one of the significant frontier projects in the United States: it was a colonization by government agents on a par with that by military posts and roads. The Great Firefight was a bold undertaking, almost heroic in its ambitions. When the era concluded in the early 1960s, the torch had changed hands from folk practitioners and private landholdings to state foresters on a public domain. Ironically, as the transfer proceeded, attitudes inverted, such that the general public came to renounce fire while foresters came, at first reluctantly, then ardently, to embrace it.

There are many ways to retell this history. One useful device is to divide that long chronicle into segments.

There is, first, a long period of roughly 60 years, during which an industrial fire regime slowly replaced a rural one through processes of technological substitution while the rural population slowly fell as a proportion of the nation's whole. The policy of fire exclusion persisted until its economic cost

[11a]

Figure 11. The preindustrial firefight. The preindustrial firefight used rudimentary techniques. Crews hacked fire lines (a) at Mount Hood, Oregon, in 1910. On the Nantahala National Forest and elsewhere (b), tools were prepositioned in locked sheds for rangers to collect on their way to a fire. Smokechasers (c) attacked a smoldering fire by first felling a small tree into the smoking one; one man shinnied up and stood on a branch (upper left) while another brought him a tin of water, seen halfway up the trunk. Probably no better testimony exists to the fervor with which early rangers grappled with fire.

[11b]

[11c]

and ecological legacy demanded a reintroduction of fire. This reckoning occurred on the public domain between the mid-1960s and the late 1970s.

There is a second wave of major policy reform, each lasting 30 or so years. The first begins in 1905 with the Transfer Act that established the U.S. Forest Service and gave it responsibility for the forest reserves. The second starts in 1935, when the Forest Service decreed that every fire should be suppressed by 10 a.m. the day following its initial report. The so-called 10 a.m. policy was quickly adopted by the other federal land management agencies, but by the mid-1960s, criticisms and new legislation forced agencies, beginning with the National Park Service, to renounce it. Despite accumulating criticism, the complete unraveling of the 10 a.m. policy, the philosophy behind it, and the political monopoly that held it in place took another 30 years. In 1995, a common federal policy was announced that required of the agencies only an "appropriate response." No longer were fire officers required to prevent or suppress every fire: some they might let roam and some they should start. In 2008 "appropriate management response" became formal doctrine and completed the repudiation of the 10 a.m. policy.

Finally, there are shorter, roughly 20-year waves, each dominated by some particular fire problem. From 1910's Big Blowup to 1930 it was the frontier fire, in which agencies had to cope with the lingering presence of settlement, from enthusiasm for light burning to the absence of fire towers and telephones to the uncertainties of new, unsettled agencies.

From the early 1930s to the postwar era, the driving concern was fire in the backcountry, not only in remote mountainous areas but also in lands remote in time, as it were—abused and abandoned lands for which there was no near-term economic payback. Thanks to the Franklin Roosevelt administration, almost overnight, fire protection burst into the backcountry as a universal presence. The Civilian Conservation Corps (CCC) created a civilian army committed largely to planting trees and fighting fires.

From the early 1950s to 1970 or so, an obsession with controlling large fires—"mass fires," as they became known—dominated thinking. The old military analogy revived, and war surplus equipment allowed fire agencies to mechanize quickly.

Then came the upheaval, fire's great cultural revolution. National discussion about fire in wilderness was prompted by new thinking about fire's ecology, public enthusiasm for wilderness (codified in the 1964 Wilderness Act), and mounting criticism about fire protection's disconnection from land management. This era reached a climax in the Yellowstone fires of 1988.

Since approximately 1990, the issue of flame and exurban sprawl, or what the agencies have chosen to call the wildland-urban interface, has commanded

[12b]

[12c]

much of the backcountry remained remote; too many fire officers hesitated when attacking such burns, fearful of expense and difficulty; too many fires were left to loiter, some fraction of which would blow up. The Forest Service had faltered, he insisted, because it had not really committed fully to fire control. The proper policy was to establish unequivocal goals and then apply the means necessary to meet them.

The third group disliked the emerging polarity between doing nothing and doing everything. Yet, as participants continued their discussions, it became clear that the middle way solved nothing. Unless all fires were attacked, some would become large and overwhelm the good results of earlier successes, and their expenses would vaporize any savings from a tempered program of suppression. The only way to contain the large fires that ran up costs and damages was to control every small fire. In retrospect, it is clear that the committee came to the conclusions it did because of the kind of questions it asked. It framed the issue in terms that compelled a particular conclusion.

Meanwhile, other criticisms bubbled up. Most came from professionals outside forestry, such as wildlife biologists, range scientists, and prairie restorationists. All saw value in burning, and most worried about the effects of favoring trees over grass, which is what the suppression program did. By now even a few critics were emerging from within forestry, notably H. H. Chapman at Yale, and they were concerned about regeneration in the prime timber species of the Southeast. They saw evidence aplenty that longleaf and loblolly pines, particularly, were well adapted to the kinds of woods burning that had traditionally swept the piney woods, and they saw little evidence of regeneration after fire exclusion.

Worse, at the 1935 Society of American Foresters meeting, they learned that even the Forest Service had discovered these facts through its experimental plots and had suppressed the results because it considered them both scientifically flawed and politically incendiary. Those experimental results couldn't be right—every tenet of academic forestry held otherwise; the plots must have been mishandled. Besides, fire protection was what joined the southern states to the national forestry agenda. Without grants for fire control, those bureaus might wither away, and the credibility of forestry suffer. The Forest Service had committed to an aggressive program to end southern woods burning—had even hired a psychologist to explain why rural folk insisted that their traditional fires did good when any rational observer knew they harmed the land and contributed to a stifling poverty. For all those reasons, the fire ecology data had been ignored, and when the studies were revealed at the 1935 meeting, critics were furious. This was, as one put it, the "first time we have been told the truth."

[13a]

Figure 13. Extending fire protection into the backcountry. Almost overnight, the Civilian Conservation Corps made firefighting possible throughout the national forests and parks. The fire cache at the North Rim of the Grand Canyon (a) opened in 1936 and remains in use today. CCC enrollees constructed many fuel breaks; the largest, the Ponderosa Way (b), spanned 650 miles across the entire western front of the Sierra Nevada.

One final critique appeared, again from Elers Koch, who objected to what he regarded as a mindless construction program in the name of fire protection. Even the Lolo Pass through Montana's Bitterroot Mountains, whose steep difficulty had once tested Lewis and Clark, now had a road, guard stations, and a lookout, deemed essential to fire suppression. A fire program in the backcountry, which Koch considered ineffective, was not only failing to protect forests but also desecrating the land's wilderness character and the nation's natural and cultural heritage. Some values, he insisted, were greater than fire control.

All these concerns landed on the desk of Gus Silcox. The chief had been the second-in-command during the 1910 blowout and had afterward written that the lesson of the Great Fires was that they were controllable; all it took was sufficient resources and political will to catch fires while they were tiny. Now, thanks to the CCC, those resources were on hand, and the veteran of 1910 had lost

[13b]

nothing of his determination to prevent such an event from ever recurring. Just prior to fire season, he announced what became known as the forester's policy or the 10 a.m. policy: the goal of fire control would be to contain every fire by 10 a.m. the morning following its report, or failing that, to contain it by 10 a.m. the next day, ad infinitum. There would be only one standard, one administrative yardstick for measuring success, and even though it was presented as an "experiment on a continental scale," the policy acquired the authority of an agency mandate. For the next 35 years the 10 a.m. policy dominated fire administration on federal lands.

A MILITARY MODEL

The 10 a.m. policy seemed entirely appropriate as the country entered into World War II. Already by 1939 it inspired the Forest Service to establish two specialty fire crews, the smokejumpers for attacking small fires in the backcountry, and the 40-man crew, self-styled "shock troops" for use in large campaigns. Even as the agencies saw their young men leave for military service, the country remilitarized fire control and nationalized its concerns.

OUR CARELESSNESS
Their Secret Weapon
PREVENT FOREST FIRES

SMOKEY SAYS—
Care will prevent
9 out of 10 woods fires!

[14a]   [14b]

Figure 14. Nationalizing fire prevention. World War II engendered a truly national firefighting program. In 1942 the Wartime Advertising Council promoted fire protection as defense and equated careless fire with sabotage (a). After the movie *Bambi* was released, the program used Bambi and friends, but when Disney Studios refused to release rights, it needed a cartoon substitute. Smokey Bear (b) appeared in 1944. A bear cub found after a New Mexico fire in 1950 became "Little Smokey."

World War II was itself a war rife with incendiaries—from flamethrowers to napalm to aerial firebombs to the atomic bomb. In 1942, alarmed over prospective Japanese incendiary attacks on the West Coast, the Wartime Advertising Council launched a program of hard-core propaganda that identified careless fire with enemy fire. The following year the Ad Council experimented with Bambi, and in 1944 it created Smokey Bear. The next war, authorities warned, would be a fire war on an even grander scale. Civil Defense had to embrace fire protection. Fire protection had to stop big fires.

The postwar adjustments did not fully occur until after the Korean conflict ended in 1953. By then a cold war on fire was underway. War surplus hardware poured into the fire agencies, not only the federal bureaus but their state cooperators. Several equipment development centers arose to beat swords into shovels, and in 1954 Operation Firestop explored the prospects for updating

[15a]

Figure 15. The cold war on fire. With a military example, and often with Department of Defense funds, fire protection established itself as part of national security. Jeeps, parachutes, and other war surplus equipment released after the Korean War mechanized fire protection (a). Helicopters were used as early as 1946, and air tankers, in 1956. Science was also mobilized. The wind tunnel at the Intermountain Fire Science Lab in Missoula (b) is being observed by Dick Rothermel, who developed a fire behavior model in 1972.

the apparatus of firefighting. Helicopters were used in 1946; aerial tankers were operational in 1956. Meanwhile, on the model of wartime mobilization, science too scaled up, boosted with funding from the Office of Civil Defense and the Department of Defense, and linked with other research bodies through a National Academy of Sciences Committee on Fire Research. In short order, a suite of national fire labs gathered unto themselves the major projects, nudging fire science away from forestry and into the physical chemistry of combustion, the physics of fire behavior, and administration by fire danger rating. The labs

[15b]

had a regional logic as well as national themes, with facilities at Macon, Georgia; Missoula, Montana; and Riverside, California. The war on fire had gone national. It would emulate the cold war in its appeal to science for understanding, technology for tools, and a sense of national security for justification.

There were a few checks that sought a better balance. The South remained stubbornly exceptional. In 1943, with manpower down but the tractor plow ready for fireline duty, the Forest Service approved controlled burning in the southern forests to help decrease fuels. There was simply no other practical alternative to keep the "rough" under wraps; to let it erupt in profusion was to invite fires to leap from the understory into the crowns in uncontrollable fury. Southern forestry began to adapt rural fire to its own ends, and as commercial forests (for pulp) claimed larger proportions of the landscape, routine burning became the norm. Academic forestry was slow to accept the practice: burning was reluctantly regarded as the temporary best of a bad set of options. But controlled flame was on the ground, and as critics sought to replace the 10 a.m. policy with something smarter, it comes as no surprise that they found voices and examples from the Southeast. The region remains to this day the premier scene for prescribed fire.

By the mid-1960s the Forest Service had more or less achieved its original ambitions. It had a virtual monopoly over firefighting resources, over fire research, and over policy. Practically every institution and exercise that involved free-burning fire had it as a member, if not a master; the last state, Arizona, signed onto the Clarke-McNary program in 1965. Even the climate cooperated, and it seemed only a matter of time, of doing more of the same, before the big fires ceased altogether and America's wildland fire scene would resemble its urban counterpart. Fire—the ancestral scourge of forestry—was on a timetable to expire.

CHAPTER FIVE

Fire's Cultural Revolution

The reformation began, like all great revolutions, with a change in values. A small group saw fire and its place on the land differently, and as the 1960s progressed, those ideas remade institutions. In 1961 the 10 a.m. policy and the Forest Service's fire monopoly were unchallenged. By 1969 significant elements of the fire community, particularly researchers and those who were not foresters, were convinced that fire belonged on many lands, that prescribed burning was useful and necessary, and that forestry was inadequate to the task. The costs of fire exclusion had grown too great. In rapid succession the old order fell apart and a new one sought to integrate the competing philosophies and factions that had brought it down.

In retrospect it was an era poised for change, and the first boulder that rolled over the cliff quickly set off a landslide. In 1962 the privately endowed Tall Timbers Research Station north of Tallahassee commenced a series of annual conferences on fire ecology. The next year a report commissioned by the secretary of the Interior on elk at Yellowstone evolved into a bold recommendation to recharter the principles by which the National Park Service governed its natural areas. The Leopold Report, named after its chair, A. Starker Leopold, son of Aldo, identified fire's restoration as a critical task for making the parks "vignettes of Primitive America." In 1964 the Wilderness Act gave enthusiasm for the wild a statutory reality and challenged all the federal agencies.

The war on fire soon confronted multiple insurrections. One came from a public aroused by environmental concerns, inspired by wilderness, and skeptical of officials and technical experts. The authorities were proving wrong on atomic fallout, pesticides, and other topics; there was no reason to believe that the self-designated authorities on fire were any more correct or that the values they proclaimed were either necessary or suitable. Forestry, in particular, presented to the public an unhelpful alloy of professional arrogance and political

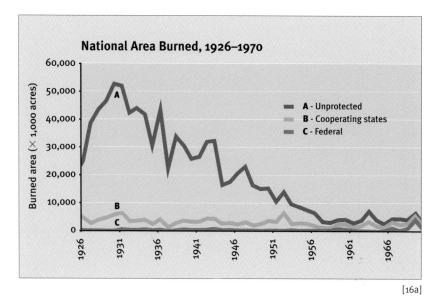

[16a]

Figure 16. Fire suppression: The good, the bad, and the ugly. Creating a first-order fire protection system can yield impressive results, at least initially. Burned acreage (a) plummeted during the 1900s as more and more land was brought under formal protection. While the number of acres burned on federal and state managed forests remained relatively constant from the 1920s to 2000, the largest decreases in acres burned were due to increased suppression efforts on private and other areas not yet protected. Control of fire over time led to the build-up of fuels on the ground, which, in part, led to increases in costs (b) to keep the lid on fire. By the 1960s the ecological costs of removing fire on this scale were becoming apparent, from stagnating biotas to stockpiling fuels. More firefighters died from mechanical accidents (c) than flame, but the cost was real and, after the 1994 season, declared unacceptable.

stubbornness, and when challenged, it appealed to authority, tending to dismiss its new critics as it previously had Piute forestry. As the United States sank into a quagmire in Vietnam, the military metaphor that had been the basis for fire suppression became a liability.

Another protest came from scientists newly sensitive to fire's ecological services. The older concept that fire interrupted the progress of landscapes toward stability gave way to an understanding that fire was integral to pathways of energy and the cycling of nutrients. It even appeared that some species adapted to fire not simply to protect themselves from flame but to promote their own survival; without the proper fire regime, they became maladapted. Clearly, fire's exclusion could be as disruptive as its application. The crisis that afflicted many

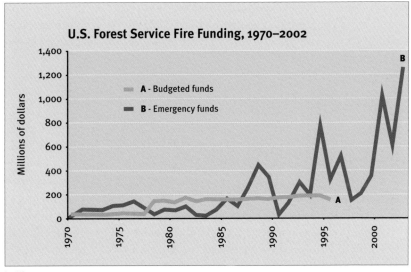

[16b]

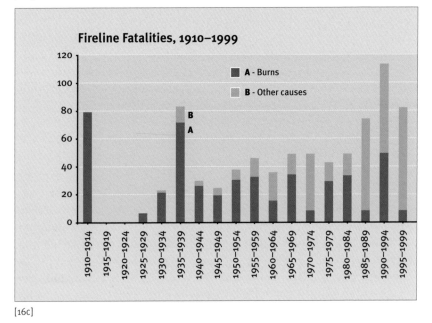

[16c]

fire-prone public lands stemmed not from free-ranging wildfire but from fire's aggressive and increasingly mechanized suppression.

It became difficult to argue that natural fire did not belong in natural areas, and if it was good there, why shouldn't its goodness be distributed widely through the proxy of prescribed burning? Nature, too, joined the insurgency; the light burners who predicted that fire exclusion would ripple through woods in unhealthy ways were proving correct. Very quickly, astonishingly so, intellectual opposition collapsed.

INDEPENDENT VOICES

The revolutionaries had a common cause in that they had a common foe. They all detested the paramilitary swagger and waste of fire control and scorned its justifications. It had become, they said, a law unto itself, divorced from the purposes of protected reserves and insulated by its reliance on off-budget emergency funding. It wasn't enough, they insisted, to fight bad fires. Agencies had to reinstate good ones as well.

The protest had two poles, one on each coast. One lay in Florida, at the Tall Timbers Research Station. This group and its charismatic director, Ed Komarek, promoted a heritage of fire as used on private land, of fire as a historical presence and cultural artifact, and of fire as a means to promote biotic assets, whether bobwhite quail or open-woods cattle. Its poster child was the longleaf pine. The other pole centered on the national parks of the West, particularly the Sierra Nevada, and had its intellectual anchor at the University of California–Berkeley. It focused mostly on public lands and found for its prophets wildlife professor Starker Leopold and range professor Harold Biswell. Its poster child was the giant sequoia. The Florida protestors wanted fire in the hands of people; the California group, as far as possible left to nature.

The Tall Timbers fire ecology conferences offered a forum outside the control of forestry and the Forest Service (or even the federal government). It published counterexamples from around the world, all demonstrating the values of proper burning and the costly folly of fire exclusion. The Wilderness Act, because it protected areas within the holdings of the federal land management agencies, provided a legal basis for Forest Service critics to question foresters' strategies and methods of suppression. By stonewalling efforts to move lands into wilderness, moreover, the agency soon alienated a public sensitive to environmental themes. Over the winter of 1967–68 the National Park Service reformed its administrative guidelines, largely along the lines proposed by the Leopold Report. It shed the 10 a.m. policy and proposed instead to encourage fire—corrective burns where necessary, and natural fire where possible. To do this, it decided it could not rely on the Forest Service for fire crews or research any more than for policy and set about establishing an autonomous organization to promote its special ambitions. The old monopoly cracked.

The Forest Service took another decade to bring itself into alignment. It experimented with fire experiments in wilderness areas by 1972, and in 1974 it announced the full integration of fire, in principle, with land management and the various planned uses of the national forests. Interestingly, the announcement was made at a Tall Timbers fire ecology conference. In 1977–79 a series of reforms sought to restructure the cooperative fire programs under the Clarke-McNary Act (the states could do the work on their own), replace reliance on emergency funds with a more traditional budget, and abolish the 10 a.m. policy completely in favor of a policy of fire by prescription. The other federal land agencies, all in the Department of the Interior, were redefining themselves along the same lines, pushed by wilderness advocates and pulled by ecologists' enthusiasm for burning. Here was American fire's Second Great Awakening.

New ideas forced new institutions. Even as fire policies were diverging to better reflect distinctive organizational missions, the practical demands of fire suppression, limited resources, and the collective identity of the fire guild were bringing agencies together in the field. What the Forest Service had for decades held together under its own authority broke apart and had to be reassembled under new auspices. A year after the Park Service announced its separate policy, the Boise Interagency Fire Center (later, National Interagency Fire Center) opened in Boise, Idaho, as a collective warehouse and dispatching facility. In 1976 the National Wildfire Coordinating Group, bringing together all the federal agencies along with a representative of the National Association of State Foresters, commenced as an advisory board to promote common equipment, training, vocabulary, and other practices to support a doctrine of "total mobility": any fire equipment, crews, or commanders attached to one agency could be interchanged with another's. Still further collaboration occurred with the development of the National Incident Management System, which provided a common organizational structure for fire emergencies, or for that matter, any emergency. The scheme has since become international.

Fire's great cultural revolution had taken about 15 years from the first barricade to a formal overthrow of policy. But ideas are easier than implementation, and institutional reorganization simpler than change in behavior on the ground. Even with benevolent weather, the shift to fire by prescription was awkward, tentative, error prone, and frustratingly incomplete. Experimentation with novel tactics and concepts conflicted with bureaucratic instincts to impose order. And there were other glitches, such as court-ordered affirmative action programs that disrupted the traditional training flow of personnel and their values.

On the ground, there were successes and failures. The successes, while real, seem largely symbolic in terms of area treated, and most occurred in places either remote from public inspection or still fluffy with grasses and hence amenable to relatively direct burning. The failures took several forms. They included pre-scribed fires that went wild, natural fires that bolted beyond their wilderness borders, burns that fizzled without having done the ecological work required, and burns that smoked-in roads and cities. After each such experience, standard procedures were quietly rewritten to prevent repetitions, so far as possible.

It became increasingly apparent that putting fire back onto the land was dif-ferent than taking it out, and far trickier. It was not enough to stop suppressing fire, or to light instead of fight. Over the years the environment for burning had changed. Combustibles had piled up and rearranged themselves such that fires, whether from lightning or torch, behaved differently than they would have in presettlement times. Some species required special protection, some had invaded and promised to shift fire regimes. Surrounding lands were no longer agricul-tural but urban, and fires were less likely to enter reserves from neighboring landscapes than to race the other way. Air-quality boards restricted the smoke that burns were allowed to produce. Fire's physical, social, and legal environ-ment had altered. If the default setting for fire management remained suppression, that was only because wildfire often appeared as a crisis that demanded a response.

EARLY DAYS

For a handful of years amid the attempted reformation, the weather was mild and the national fire scene quiet. By the early 1980s burned area shriveled, while climatologists warned of an impending ice age, and the Great Salt Lake, bloated with rains, overflowed its levees. Then the West commenced a long drought. Some wildfires reappeared in 1986; stubborn burns prolonged fires in northern California into the so-called Siege of '87; and a scourge of burns lashed Yellowstone National Park through the long summer of 1988. By now climatologists had reversed themselves and were warning not of nuclear winters but of greenhouse summers. Humanity was turning Earth into a Crock-Pot of irreversible warm-ing. Large fires in Amazonia and Yellowstone, both heavily televised, were presented as the pilot flames of a coming apocalypse.

The Yellowstone fires were a cultural moment as the fires stayed in headlines and on TV news for weeks. They alerted the public that traditional fire man-agement had ended and a new order was at hand. They advertised, as perhaps no other event could, the recognition that fire belonged in many ecosystems as fully as wolves and grizzlies. They educated the American public (and much of the world) to the might and majesty of free-burning fire in Earth's ecology. But they were equally a discussion that *didn't* happen. They focused people's atten-tion on the effect of fire in the environment rather than on how such fires ought to occur and how they should be managed. The fire community united to

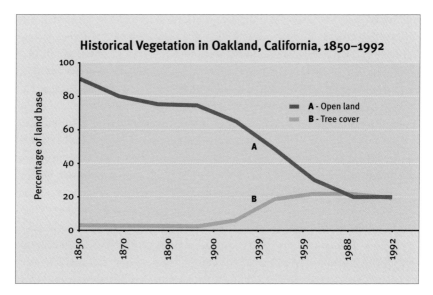

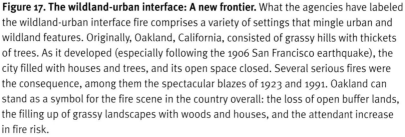

Figure 17. The wildland-urban interface: A new frontier. What the agencies have labeled the wildland-urban interface fire comprises a variety of settings that mingle urban and wildland features. Originally, Oakland, California, consisted of grassy hills with thickets of trees. As it developed (especially following the 1906 San Francisco earthquake), the city filled with houses and trees, and its open space closed. Several serious fires were the consequence, among them the spectacular blazes of 1923 and 1991. Oakland can stand as a symbol for the fire scene in the country overall: the loss of open buffer lands, the filling up of grassy landscapes with woods and houses, and the attendant increase in fire risk.

support fire's necessity. They celebrated the triumph of the new ideology and deflected criticism about practical management—at what cost, by what means, and through what social compact fire's restoration should occur. A critic might well argue that Yellowstone's fires were equally an ecological benefit and an administrative debacle.

The fires of '88 closed out an era dominated by wilderness fire, institutional experimentation, and relatively benign climatic conditions. After a several-year pause, big fires returned, more savage than ever, and the media, sensitized to fire's ability to capture viewers' attention, prepared to broadcast them as part of an annual cycle of disasters and spectacles, like hurricanes and volcanic eruptions. For a decade, beginning and ending in Southern California, from the 1993 Malibu fires to the 2003 Halloween fire bust, fires became celebrities. The biggest fires happened in even years, helpfully contributing photo ops for national elections. The 1994 and 2000 seasons especially shocked the fire establishment.

The 1994 season racked up the highest burned area in half a century, saw suppression costs exceed a billion dollars for the first time, and killed 31 fire-fighters, including 14 at the South Canyon fire outside Glenwood Springs, Colorado. Suddenly the full costs of America's fire policy were visible for all to see: the ecological costs, the economic costs, the costs in lives. Yellowstone had celebrated fire's value; South Canyon presented fire's bills due.

What catalyzed the transition, however, was the posthumous 1992 publica-tion of a meditation on the 1949 Mann Gulch fire by Norman Maclean, the well-known author of *A River Runs through It*. His *Young Men and Fire* con-nected American fire with America's intellectual and literary classes in ways that had not been evident since 1910. Wildfire no longer appeared as some bit of western exotica like a grizzly bear attack but something through which one could address even fundamental questions of existence. At Yellowstone, fire manage-ment made an ecological connection between burning and valued landscapes; at South Canyon, it made a cultural connection between firefighting and a broader society. A new wave of reforms swept through the federal fire establishment, cul-minating in the adoption of a common agency policy in December 1995.

The biennial busts continued in 1996, scattered throughout the West, and in 1998, scorching Florida and forcing more than 100,000 people to evacuate—a veritable fire hurricane. The climax came, however, in 2000 as the fire commu-nity seemed unable to either fight fires or light them. Wildfire returned to the Northern Rockies with implacable force, as though 90 years of exhaustive endeavor since the Big Blowup had meant nothing. But the onset of the season witnessed two spectacular failures in prescribed fire, both by the National Park Service. The Outlet fire set at the Grand Canyon blew up and forced the evacu-ation of the North Rim; the next day, a failed prescribed burn ignited at Bandelier National Monument bolted loose and, renamed the Cerro Grande fire, scoured out Los Alamos, New Mexico, the site of a national weapons lab. The public saw a scene in which nuclear weaponry itself was challenged by a much older firepower.

The fire community admitted publicly that it could not do what it wished. It still believed that it knew what to do but feared that conditions—both environ-mental and political, the worsening climate of the West and a souring climate of public opinion—would impede its mission. In 1998 Congress endowed the Joint Fire Science Program to rekindle a smoldering program of fire research. By year's end the National Fire Plan had been approved to introduce the reforms and fund-ing deemed necessary to suppress bad fires and support good ones. Then the 2002 season shifted the shockwaves to the Southwest. The largest fires recorded for Arizona and Colorado, powered by the worst drought in a millennium and the accumulated abuses of a century of land use, mesmerized the public. The

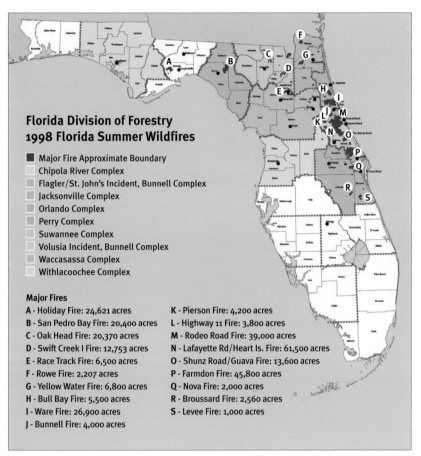

Florida Division of Forestry
1998 Florida Summer Wildfires

- ■ Major Fire Approximate Boundary
- ☐ Chipola River Complex
- ☐ Flagler/St. John's Incident, Bunnell Complex
- ☐ Jacksonville Complex
- ☐ Orlando Complex
- ☐ Perry Complex
- ☐ Suwannee Complex
- ☐ Volusia Incident, Bunnell Complex
- ☐ Waccasassa Complex
- ☐ Withlacoochee Complex

Major Fires

A - Holiday Fire: 24,621 acres
B - San Pedro Bay Fire: 20,400 acres
C - Oak Head Fire: 20,370 acres
D - Swift Creek I Fire: 12,753 acres
E - Race Track Fire: 6,500 acres
F - Rowe Fire: 2,207 acres
G - Yellow Water Fire: 6,800 acres
H - Bull Bay Fire: 5,500 acres
I - Ware Fire: 26,900 acres
J - Bunnell Fire: 4,000 acres
K - Pierson Fire: 4,200 acres
L - Highway 11 Fire: 3,800 acres
M - Rodeo Road Fire: 39,000 acres
N - Lafayette Rd/Heart Is. Fire: 61,500 acres
O - Shunz Road/Guava Fire: 13,600 acres
P - Farmdon Fire: 45,800 acres
Q - Nova Fire: 2,000 acres
R - Broussard Fire: 2,560 acres
S - Levee Fire: 1,000 acres

[18a]

Figure 18. Flames in Florida. Large wildfires swept through Florida in 1998 (a) and again in 2007. The nearly 500,000 acres burned by 2,282 fires in 1998 does not include the extensive areas burned annually by prescribed fire. All jurisdictions in the state, from private and county to state and federal agencies, conduct prescribed burns (b), and Florida has established itself as a major center for fire management. Nevertheless, the area deliberately burned remains well below historical levels: an early state forester proclaimed that 105 percent of the state had burned over the previous year, the result of ranchers' burning in the spring and then reburning much of that land again in the fall.

2003 season ended with a plague of fires in Southern California, culminating in the Cedar fire that abraded Scripps Ranch outside San Diego. At the end of that year, the Healthy Forest Restoration Act of 2003, which sought, among other matters, to institute a modest program of thinning and understory clearing to unclog clotted woods, was signed into law.

[18b]

OUT OF THE ASHES

By now there was widespread appreciation that something was deeply amiss in America's fire scene. Each year the figures for burned area, costs, and damages seemed to rise. As the federal budget plummeted into record deficits, Congress demanded that the Forest Service, in particular, pay for fires out of its authorized funds, not by unbudgeted deficits of its own, now that fire costs commanded 50 to 60 percent of its regular allotment. Acreage from prescribed fire nudged upward, but doubling the area burned had required a fivefold increase in costs. Firefighters continued to die. Fire officers were held liable for failures, and partly because of threatened lawsuits for wrongful deaths (or even criminal negligence), they were urged to take out personal liability insurance. Several natural fires that were allowed to burn erupted and wiped out decades of dedicated site protection, not simply of timber (since logging was often shut down) but of habitat and nesting sites for protected species such as the northern goshawk and Mexican spotted owl.

Ironically, the largest fires in Arizona, New Mexico, and Colorado had all been started by people associated with the fire community—the Rodeo-Chediski fire by an Apache Indian looking to be hired to fight it, the Cerro Grande fire by a National Park Service prescribed fire crew, and the Haywood fire by a Forest Service prevention technician supposedly hired to forestall accidental ignitions. In 2007, when a 30-year veteran of the Coconino National Forest, Van Bateman,

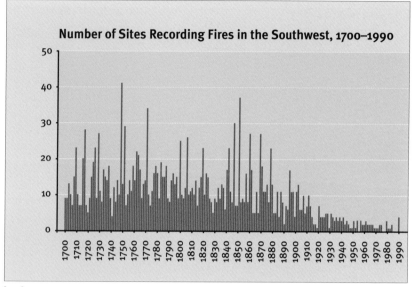

Figure 19. Southwest summary. The story of the Southwest is generally a story of the West. The number of sites in the region that experienced fires (a), as determined by tree scarring, tracks a long-term pattern of burning tied to drought cycles and mixed ignition sources. The sudden collapse in the late 19th century was due to overgrazing, which removed small combustibles. However, large fires returned to Arizona forests in the late 1900s (b), though with an intensity higher than that witnessed earlier in the century.

was convicted of setting fires, his defense was that he was only doing what bureaucracy had made too cumbersome. He was getting fire back on the land.

It was a bizarre incident, but one that might stand as a concluding statement of where the revolution had gone. Over the course of 40 years, the fire community had completely reversed its philosophy toward fire and reorganized both its policies and its institutions to reflect that reformation. Although the fire community, and its allies among environmentalists, continued to argue that the choice was between fire's suppression and fire's restoration, the real issue was the various ways and means to promote the fires we wanted while preventing those we didn't. The reinstatement of fire on the landscape wasn't happening as easily, cheaply, safely, or benevolently as the prophets had forecast. Rather, with each year it became more complicated, more expensive, more hazardous, and more prone to unexpected consequences. The revolutionaries seemed poised to replace one fire fundamentalism, the 10 a.m. policy, with another—one that said fire had to get back on the land as soon as possible, one that could tolerate little dissent

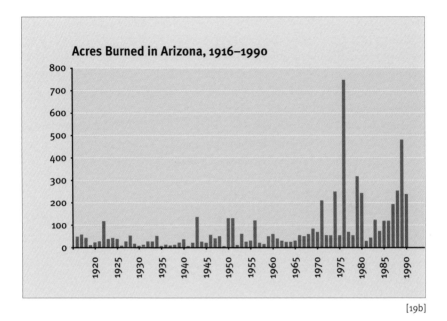

Acres Burned in Arizona, 1916–1990

[19b]

from the new consensus. Few noted that for some agencies, the duration of the new fire era exceeded the tenure of the 10 a.m. policy, and that for America's fire history overall, the length of the new era was fast approaching the longevity of the old one.

The clarity that had animated criticisms—the belief that we should use fire as well as suppress it—had evaporated. Putting fire back into the land was not simply the reverse of taking it out. Reinstating fire did not mean simply downgrading suppression, creating a parallel organization, supplanting one bureaucratic empire with another, or replacing a bias for fighting fire with a bias for lighting it. Nature's fires might not restore nature's splendor, especially when acting on a much altered landscape and amid a changing climate. Wild fire could well mean wild landscapes without regard to biodiversity, historic conditions, or perceived naturalness. A policy based on "appropriate management response" could eliminate the simple prescriptions of the 10 a.m. policy but could not easily define *appropriate*. It could prevent fire officers from mindlessly attacking every fire, but it did not specify what they *should* do. Its success would depend entirely on the adequacy of local knowledge, skill, and planning.

And there was less tolerance for experimentation. The once-rural landscapes that had helped buffer wildfire were filling up with houses. Even the period of grace that wet years had granted the early revolutionary era was gone. Overstocked fuels did not decompose on their own, weather did not cooperate with approved burn plans, managed fire did not magically convert degraded landscapes into quasi-pristine ones. Messed-up landscapes, it seemed, might only yield messed-up fires.

Fire's great cultural revolution had been a creature of its times. The old regime had no longer made sense to Americans, so they had removed it. But consensus about what the new order should do, and how and at what cost, was missing. There could be no return to the "continental experiment" of the 10 a.m. policy, certainly. Fire in some form belonged on fire-prone public lands. But the new era needed to mature if it was to deal with complexity, contingency, and context. Such understandings, however, are not the normal stuff of revolution.

CHAPTER SIX

Fields of Fire:
Problems and Strategies

As the new millennium unfolded, America discovered that it did not have a fire problem. It had many fire problems. For some, technical fixes were possible. For others, the uncertainties of fire had to mingle with the ambiguities of culture, and solutions would involve social choices and political compromises. There was too much unknown, too much unstable, and too much lacking in social consensus. A sympathetic observer might note that accepting this condition was the true reward of the revolution, that past efforts to distill multiple problems into a single, comprehensive solution had done more harm than good.

How to understand the choices involved? There were lands that were intrinsically fire prone, and lands that were not, and lands that could be made more so. Each required different strategies for fire management. There were lands that belonged in the public domain, lands that remained in private hands, and lands in both that were rapidly changing their character. These, too, required distinct approaches.

STRATEGIC OPTIONS FOR PUBLIC LANDS

The political controversies have swirled most fiercely around those lands that are both public and fire prone. A century's experience suggests four general approaches are possible. Each has its advantages and liabilities.

1. Leave alone. One option is to stand aside. Let nature's fires burn nature's landscapes and make whatever correctives are necessary to disrupted ecosystems. This strategy has undergone several changes in name as well as practice. In the 1970s and 1980s it was called *prescribed natural fire*, meaning a fire set by natural causes but allowed to burn so long as it remained within guidelines (the

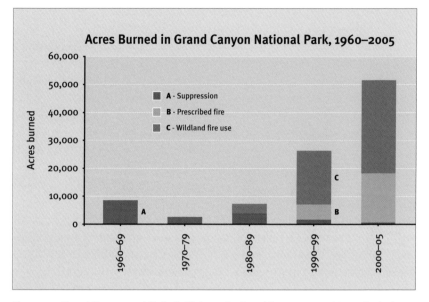

Figure 20. Grand Canyon and Kaibab Plateau. Restored fire can resemble ecological shock therapy. Significant fractions of Grand Canyon National Park and the Kaibab Plateau generally have burned within the past 20 years. A casual glance at burned area figures alone might suggest global warming as a motive force. But as this graph for Grand Canyon National Park shows, the immediate causes of the increased burning are fire management practices, which have resulted in extensive efforts at wildland fire use and prescribed burning, plus large escaped fires, all against a backdrop of regional drought. Not shown on the graph are two massive fires on the surrounding Kaibab National Forest, both of which began as wildland fire use burns, which subsequently blew up and burned over 50,000 acres each.

prescription). By the early 2000s it was known as *wildland fire use* and had become the treatment of choice for wilderness and natural reserves.

Wildland fire use avoided legal liability because lightning started the fire, and it was generally favored by environmental groups. On the downside, some of those burns would bolt. The cost of fighting such escapes could well exceed the savings enjoyed by not suppressing them earlier, might damage ecological values if they burned outside the historical range of conditions, and could threaten the political trust that allowed such practices.

2. Exclude. The second option is to exclude fire so far as possible. The drive to prevent fires from starting and to suppress those that do has always been central to fire management efforts, and it remains essential to protect economic and ecological assets. Fire suppression will not wither away. Any other strategy requires it, if only in reserve. The firefight remains the great constant of fire administration.

Yet it is clear that the firefight, at times necessary to restore order, is not the same as governance. The attempt to make fire suppression the basis of land management has only catalyzed ecological turmoil and created a problem that no amount of further firefighting can extinguish. In fire-prone landscapes the more one attempts to exclude fire, the worse the fires that do happen can become.

3. Prescribe burn. A third option is to accept that in many places fire is essential to do necessary ecological work, but people should do the burning. Prescribed fire is a compromise between wholly natural fire regimes and wholly artificial ones. It is, moreover, the historical experience of humanity up until the industrial age, and it is possible to argue that it was the elimination of humanity's burning, not the suppression of natural fires, that most altered America's landscapes and accounts for much of the decline in burned area over the 20th century. These missing fires are not simply fires that nature once set and people suppressed, but fires people had traditionally kindled and no longer do. The more recent uptick in burning may be the outcome of feral fires made possible in part by the removal of tame ones.

Prescribed burning thus seems an obvious remedy, but it requires satisfying a growing roster of concerns, including endangered species and cultural artifacts. Now, too, the atmosphere must be protected; the Environmental Protection Agency's standards for emissions of fine particles (PM2.5) set limits on wood smoke, and standards for greenhouse gases may well regulate burning not tied to explicit ecological purposes. Moreover, some fires will escape, some will not do the work expected, and some, acting on landscapes mangled for a century, will not behave as they would have historically. Unlike the case with lightning fire, the agents responsible for setting a prescribed fire that misbehaves are easily identified and can be hauled into court.

4. Redesign the landscape. The last option is to change the landscape itself—that is, to rearrange the setting so that fire, any fire, from any source, will behave as desired. In practical terms this means cropping grass, thinning and otherwise tinkering with trees and brush, and replacing flammable vegetation with less flammable plants. If this is done correctly, a wildfire can be easier to contain and a prescribed fire more likely to perform as predicted.

This of course is an ancient strategy, and it forms the basis for fire protection in cultural landscapes such as Europe's. But it means people doing things on the land; it means chain saws and wood chippers, and perhaps goats and cattle; it means busy hands. This alarms those who want people removed as far as possible from wildlands: the strategy might seem like suppression by another means. Proposals to intervene on the scale necessary can quickly escalate into political firestorms.

The conclusion seems obvious and modest: none of these strategies will work by themselves. Every place will involve a mix of several, the proportion depending on the specifics of location. Yet because these are national lands, people who live distant from them will help decide their future; in particular, urban constituencies will decide fire practices on lands far removed from their daily lives. The future, done right, will involve a long negotiation among interest groups and between them and the land. It will require patience, humility, and experience. In the end it will probably cost as much as a suppression policy.

THE FUTURE OF FIRE SUPPRESSION

Although firefighting has surrendered its primacy as the basis for fire management, suppression continues to flourish amid drought, overstocked woods, careless ignition, and the intermix fire scene. But just as there is no longer one goal for suppression (control by 10 a.m.), so there is no single means to do it. Fire suppression has become more powerful, better informed, gentler, more flexible, and vastly more costly.

Fire suppression has always inflicted its own damages. In the past these were dismissed as trivial in comparison with the damages wrought by wildfire. Yet as the fire community came to appreciate the benefits of free-burning fires, it has fretted over the environmental costs of fighting them—the damages from the forests felled and soils overturned to make firelines, the firelines that erode into ravines, the chemical retardants that both fertilize ash and pollute waterways. If fire is to serve larger land management, then firefighting needs to reform its practices in the field, and it has. Light-on-the-land tactics have become more common in the backcountry; local resource specialists (and sometimes archaeologists) help site firelines to avoid destroying artifacts, critical habitats, and nesting sites; bulldozers are often prohibited in sensitive areas without written authorization at secretarial levels; postfire rehabilitation includes such measures as installing water bars for firelines and treating disturbed soils. Burned Area Emergency Response teams now move in as incident command teams move out. Even wars have rules of engagement, and the war on fire has accepted environmentally friendly accords that restrict what it might do and how.

After the 1994 disasters, fire suppression also became more cautious about where and why it put firefighters at risk. These choices (and reviews of agency actions) are no longer left solely to agency discretion; the Occupational Safety and Health Administration investigates fireline fatalities as workplace accidents and has issued citations for violations. Supervisors at the 2001 Thirtymile fire in Washington, where 14 firefighters and two civilians were killed, were even charged with negligent homicide. The entire culture of firefighting has slowly changed, with each new rash of fatalities leading to further cautions, threats, and the requirement to justify any loss of life.

The result is that major fires may not be attacked directly; instead, crews exploit advantages in the terrain and pick points for making stands, often some distance from the existing fire front. A fire that might once have been hit and held to a quarter acre in the mountains might now be allowed to burn out a whole basin framed by roads and rocky ridges, in the belief that such actions will be cheaper, safer, and more ecologically benign. Such decisions can, in practice, resemble wildland fire use, although they remain "suppression" in name. These changes in strategy help explain the recent increases in burned area.

PUBLIC LANDS, PRIVATE LANDS

A map of big fires today more and more coincides with public lands. There are some exceptions, such as the grassland fires that have struck Oklahoma and Texas, although even here public lands and private lands committed to nature protection feature in the mix. Overall, however, private lands remain sites of industrial fire, places where internal combustion replaces open burning. The scene for fire management is public wildlands, and these have undergone dramatic reform.

The great institutions for wildland fire emerged during a colonial era that reserved large public estates, assigned their administration to state-sponsored forestry bureaus, and confiscated fire from private persons and made it in effect a government monopoly. The past 60 years, however, has been a time of global *de*colonization, and the process has changed those institutions and compelled a realignment between private and public agents. This has happened around the globe. New Zealand disestablished its Forest Service; India adopted a doctrine of "social forestry" that effectively removed the larger agency save in name; Canada has provincial forestry bureaus that must negotiate land treaties with indigenous people; Australia completed a massive transfer of lands from forestry to parks. The United States has retained its Forest Service, but that agency has undergone a profound reinvention that has turned its mission into one of state-sponsored ecology and forced it to accommodate land uses very different from those established at its origin.

The reforms have blurred the lines between private and public. More and more operations on public lands are outsourced and privatized; they are done by contractors, not by agency personnel. But the reverse is also occurring: private landowners are reclaiming rights to fire. They had never lost them entirely, of course, and the loss was less from outright confiscation or statute than by steady encroachments that made burning more onerous and encouraged landowners to seek other technologies. Suburbanites ceased to burn lawns and autumn leaves; ranchers and farmers found it more difficult to get burning permits after air-quality regulations stiffened requirements; foresters had fewer occasions to burn slash. Traditional burning for grass seed in Oregon's Willamette Valley and around Spokane, Washington, has been shut down because the smoke was classified as an air pollutant. Even on pine plantations in the Southeast, long the center for private burning, the environmental and economic pressures are pushing landowners away from routine fire. Exurban sprawl has put more houses near burning woods; and ranchers and forest owners are often selling their land for housing developments, further pushing fire off the map. Where fire remains, it does so almost as a relic, like mom-and-pop corner stores in an era of megamalls and franchises.

To this trend there are two vital exceptions. One is Florida, where woods burning has remained an ancient right, legislation eases liability concerns for private landowners, and burning permits are issued with little delay. The other exception, more generally significant, concerns private lands committed to nature protection for which fire does biological work that is essential to the maintenance of those reserves. In such settings fire is not an economic tool but an ecological service.

Take, for example, The Nature Conservancy. Its vast prairie holdings have forced it into fire management, which it has embraced. In many respects it has become more vigorous than the federal agencies, precisely because every decision is not a political act. It trains crews, develops indices to measure success, and joins fire to the ecological considerations that are becoming the primary charge of the federal agencies. It also excels at cooperation, bringing diverse groups together to promote common concerns in fire. Because it owns its lands, it has to manage them, which makes it different from those environmental groups that seek to influence policy only on lands owned by others (or the public). On its lands, it can do, within limits, what it deems necessary, and it can negotiate easements or other arrangements with neighbors to help. It thus acquires credibility: it is not a think tank or a consortium for lawsuits but a landowner with practical experience in burning.

By the onset of the 21st century The Nature Conservancy was serving as an honest broker among groups long locked in political blood feuds. The Forest Service has turned to it to catalyze "learning groups" by which private and public lands might achieve common objectives in fire management. More astonishingly, the conservancy has carried its message overseas through a Caribbean and Latin

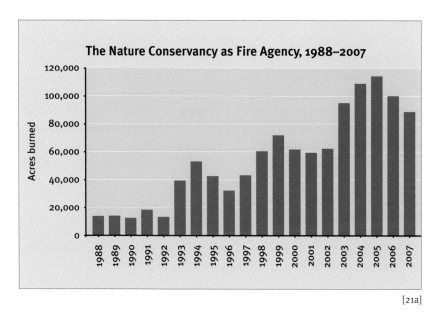

[21a]

Figure 21. The new administration of fire. The Nature Conservancy has become the best known of private owners committed to fire management. It practices prescribed burning on its own lands at escalating levels (a) while helping broker among public agencies and private groups throughout the country and exporting its experience and training capabilities to Latin America, the Caribbean, and other parts of the world. In Florida, such cooperative arrangements help boost the burning significantly (b).

[21b]

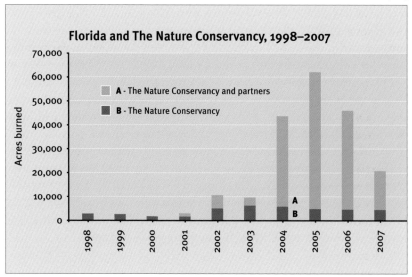

American fire initiative to promote wise fire practices in places of critical habitats and threatened species, and in alliance with the World Wildlife Fund, it is
sponsoring the Global Fire Initiative. Appropriately, it has located its fire office
at the Tall Timbers Research Station.

Thus we now have a mixture of public and private programs. The rule seems
to be this: where fire is a tool solely for reducing fuels or serving economic goals,
its use will falter, but where it does biological work for which there is really no
good alternative, it will thrive. In this regard, the critical divide is not so much
whether the land is public or private but to what purpose it is put. Function will
follow form.

WILDERNESS FIRE

The Wilderness Act of 1964 was instrumental in breaking apart what seemed
to critics an unholy alliance between forestry, the Forest Service, and fire suppression. It challenged forestry's assessment of wooded land, questioned the
agency's administration of the public domain, and discredited fire exclusion as
a management goal. It helped justify fire's reintroduction by forcing fire practices
to align with wilderness values. For roughly 20 years, from the time the National
Park Service renounced the 10 a.m. policy in 1968 to the Yellowstone fires of
1988, wilderness fire dominated the concerns of fire management.

But while the dominance of wilderness fire has faded, the actual means of
managing fire in legal wilderness and nature reserves endures. It remains, for
many reasons, beyond simple technical solutions. The definition of wilderness is
a cultural one, which means that even when people allow seemingly natural
processes to operate, that deferral still reflects cultural choices. The preferred
method for fire's restoration, wildland fire use, may not yield the expected biological results, or the fire may not stay within its borders. A fire operating on
decades of fuels accumulated because of overgrazing and suppression is not a
natural fire. Some wilderness units are small (as little as 5,000 acres), tiny enough
that a wildfire from outside could engulf the entire reserve in an afternoon. Some
such sites need crown fires for their ecological machinery to work properly. Smoke
from long-lingering burns can affect air quality and even public health.

The paradox of wilderness fire is that it will require active management, even
when some fraction of the needed work can be outsourced to nature. Probably,
too, it will require prescribed crown fires. The idea seems improbable only because,
in nature, high-intensity fires are also large-area fires. The two properties can be
separated, however, and a site burned in stand-replacing patches.

INTERMIX FIRE

What succeeded wilderness as an informing theme was a condition the agencies have awkwardly termed the wildland-urban interface—something that might more aptly be called an intermix fire because of the way it jumbles various landscapes. The ungainly term has some meaning in the West where public and private lands abut directly; it is less apt elsewhere, where houses sprout on former agricultural lands. The condition it describes, however, is national (and even international) in scope. It characterizes America from Florida to California, and industrial societies from France to Australia.

Since 1950 the population of the United States has doubled. The agricultural frontier, long ended, has given way to a frontier of urban settlement as housing developments splash over the once-rural scene. The new settlers migrate from urban centers, with urban values and urban expectations and, while insistent on urban services, are often reluctant to create local governments and taxes to pay for them. These communities coexist with natural hazards: on bottomlands subject to flooding, on coastal plains subject to storm surges and hurricanes, on earthquake belts, and on fire-prone landscapes. While spectacular, damage from fire is far less than that from water (one Category 4 hurricane is worth a century of wildfire). In terms of real economic losses, perhaps 85 percent of the wildfire damage is localized in California.

The 2007 season revealed how national the issue is. Georgia combined some of its worst elements with fires feeding on both fallowed woodlands and public reserves such as Okeefenokee National Wildlife Refuge. Later that summer Honolulu found its fringe aflame. The outbreaks came to a typical climax in California as Santa Ana winds drove fires into San Diego again. Still, as in most fire themes, there is a pronounced split between East and West. In the West, exurban developments arise mostly on former ranches adjacent to public land that is susceptible to wildfire. In the East, especially in the Southeast, the developments typically evolve out of commercial forests or marginal farmland. As the landscape fills in with houses and a thickening understory, fires return—only now the company tractors, bulldozers, trucks, and crews are gone, and the capacity for firefighting has weakened. In both regions the fires will continue until residents again actively shape the landscape to accommodate fire and until some public institutions for protection emerge. The resulting landscape will be exurban, neither a working rural scene nor a wholly urbanized one, but a quirky hybrid that will demand at least some level of manipulation.

The fire community early appreciated the problem and moved to create institutional alliances with the National Fire Protection Association and state forestry bureaus. The institutions' powers are limited. But they can inform, they can conduct inspections, and they can refuse to protect structures that are, in their estimation, "indefensible." Gradually, a consensus for fire protection is emerging, catalyzed by years of television footage showing burning houses. And gradually an understanding of why structures burn has identified roofing as the primary concern, followed by vegetation capable of carrying flame directly to combustible siding. Studies observe that ember attacks and small surface fires, often after the passage of the main fire front, are primary causes of ignition, as close-packed houses carry flames among themselves, not unlike a fire through dense canopies. Ironically, houses are more likely to spread fire to adjacent trees than the other way around.

One other ingredient determines vulnerability: the presence of someone to extinguish sparks. Here, the state of knowledge is out of sync with practice. The knee-jerk response is to evacuate whole communities, often at the last minute. This has two undesirable outcomes. One, it encourages fatalities, since most deaths occur during hurried flights and evacuations, and two, it leaves houses wholly unprotected. A single spark can doom an entire structure, which might then ignite adjacent houses. In principle, trained firefighters can replace homeowners, but in major events, there are never enough engines and firefighters, and firefighters have become wary of putting themselves between structures and advancing flames. By contrast, Australia has developed programs to permit residents to shelter in place, which has encouraged them to ready their houses and leaves able-bodied persons on site to swat out flames after the front has passed. The Australians have distilled their experience as follows: Houses save people, people save houses. Leave early, or prepare, stay, and defend.

The wildland-urban interface has created a special landscape, one that hybridizes urban and wildland firefighting. In the end, however, its fires remain a peculiar category of urban fire, and they will be contained by methods adapted from the urban fire services. Increasingly, the public has understood the larger issues, and either because of legislation or common sense, new housing and business developments are integrating fire protection into their designs and fire protection districts are sprouting up. The stubborn issues are the need to retrofit older communities, an awkward and expensive proposition, and to resolve the role of firefighters. If firefighters are reluctant to protect structures, or federal land agencies to offer a service that might properly belong with local communities, then residents should be trained to do so. One could even imagine a kind of fire militia that might be mobilized during crises, reminiscent of local fire associations a century before.

Despite its graphic presence in the media, the intermix issue is being managed, domesticated, and absorbed into the normal operations of fire agencies. Most new construction takes fire into consideration; more and more, the hazard resides in older exurbs in need of retrofitting. The fires still come, as they do in wilderness, but they no longer seem like an alien visitation. It is time to contemplate the next new thing.

RESTORING FIRE ON WORKING LANDSCAPES

"Restoring" fire has been the point of reform. Yet after 40 years the definition of restoration reflects the bicoastal origins of the revolution. In the West it means reinstating fire in wilderness, for which wildland fire use will likely become the treatment of choice. In the East it means getting, or keeping, fire in working landscapes, although the purpose of such places is to produce fewer traditional commodities and more ecological goods and services. The coming era promises to do for such working landscapes what an earlier generation did for wild landscapes.

The gist of the issue is that fire behaves as its setting permits, and that setting, historically constructed, cannot be ignored. No less than reintroduced wolves, fire requires a suitable habitat if it is to thrive and not threaten. Whether such a setting can be created by fire alone remains uncertain. There are some places where this seems possible, particularly if a grassy understory exists (such is the case with prairies and woody savannas). There are many other places, however, where reinserting fire has produced results opposite those predicted or desired. Fire can only do what its circumstances permit.

In this sense, restoration refers to the new regime overall, an attempt to reinstate better—typically former—conditions. As illustrations, consider the reintroduction of fire in Carolina Sandhills National Wildlife Refuge in South Carolina; around Flagstaff, Arizona; and in the Missouri Ozarks.

Carolina Sandhills was established in 1941 on lands long degraded by row crops and soil erosion. Wildfire was a problem in the early years, and portions were subject to artillery practice. Gradually, refuge officials tamed wildfires and then realized that the land needed fire of another kind—something to remove understory vegetation, a habitat requirement of a resident rare species, the red-cockaded woodpecker. They experimented with prescribed burning until an ice storm damaged the woods and made burning more difficult. Refuge managers persisted. After turning to aerial ignition by helicopter, they began to rack up acres. Today, with nearly 40 percent of the refuge burned annually, the program has reached a steady-state regime.

The reasons for fire's successful reintroduction are several. The Southeast has long been tolerant of burning, so there was little hostility among the still-rural population. Refuge officers could intervene actively, as they deemed appropriate—even poisoning oak, cutting out exotic slash pines, and using mechanical equipment. The landscape was still grassy, which always makes for

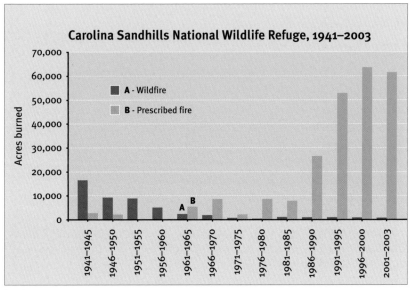

[22a]

Figure 22. Prescribed fire versus wildfire. At the Carolina Sandhills National Wildlife Refuge, site of longleaf pine and the endangered red-cockaded woodpecker, managers effected a shift in fire from wild to controlled (a). After an ice storm stalled restoration efforts, they resorted to aerial ignition to burn more than a third of the refuge annually. In the western national forests during the same era, burned area has steadily and alarmingly swollen (b). What is fascinating, however, is that the profile of the two curves is eerily similar. The refuge was able to substitute prescribed fire for wildfire; the western forests were not.

easier burning than places overrun with woody species. The red-cockaded woodpecker became officially listed as endangered, bringing the full weight of the Endangered Species Act to bear on management. As in other refuges, the administrative goal is not naturalness or wilderness but a habitat suitable for the species at risk. Despite stringent air-quality restrictions, the refuge can get fire on the land by being nimble and opportunistic and by relying on a helicopter. The acreage burned by prescription far exceeds that previously burned by wildfire at the time the refuge was created. The success of Carolina Sandhills suggests to some observers the pattern that ought to occur throughout the protected lands of the United States: a movement from wildfire to controlled fire. That such a movement has not happened may be because the refuge enjoys special circumstances.

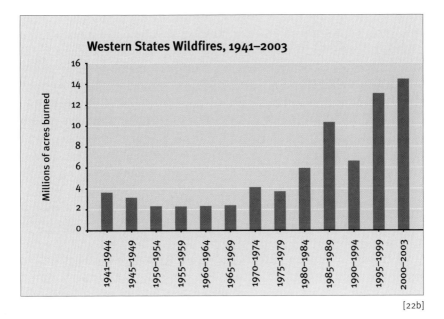

[22b]

Perhaps the best known experiment in the West is the effort to return something like a historic fire regime to the ponderosa pine forests around Flagstaff, Arizona. The mechanics of fire's exclusion in the region are reasonably well understood. The process began with massive overgrazing by sheep in the 1870s and the removal of indigenous burners, which together stopped fire dead. By the early 20th century much of the land became a forest reserve, subject to programs of active fire suppression. Throughout, here and there, logging selectively stripped out old-growth ponderosas. Over time a dense woody understory replaced the old grassy glades. Both in structure and process, the ancient ecosystem was broken and then primed to burn catastrophically. Early attempts to reverse the process by prescribed burning alone failed. More fuel sometimes remained after the burn than before; old-growth trees, girdled by slow burning around their root collar, weakened and died; high-intensity fires in places that had historically not experienced them on any scale could obliterate the biotic values that the fire was intended to promote.

Plots were then established on the Fort Valley Experimental Forest (on a site never logged) where researchers sought to reconstruct the presettlement pattern of big trees and open savannas, to eliminate grazing, and then to burn. The results were promising and led to field trials on Mount Trumbull on the Arizona Strip north of the Colorado River. Then the idea was partly codified in legislation. The Healthy Forests Restoration Act delighted some—and horrified others, and objections came in torrents. Such programs were labor intensive and hence expensive, and could succeed only if a market could be found for the small-diameter debris cleaned out of treated sites. This looked like a revived wood products industry, which is exactly what many environmental groups had sought to eliminate. It looked like the return of fussy hands and chain saws in places that some people wanted left untrammeled.

Meanwhile, partisans of landscapes not dominated by ponderosa pine worried that a model developed elsewhere (and perhaps flawed) might be exported foolishly into landscapes with quite different histories and that required other fire regimes. What did the Flagstaff model mean in oak savanna, chaparral, lodgepole pine, or cypress? If one referred only to the prescriptions themselves, nothing. But if one recognized in the experience the value of community engagement and the complexity of restoring fire, then everything. The Flagstaff model would not be cloned, but it might spawn a hundred prescriptions, each specific to particular areas.

The outcome has been mixed. Mostly, the Flagstaff program has stalled, restricted to belts around the wildland-urban interface. The primary reason is that, unlike the case with Carolina Sandhills, the Flagstaff model rammed into wilderness. Proponents of wild landscapes did not want meddling hands and wood chippers: they believed that nature could resolve its fire issues better than people could. It was the wild, not particular habitats or species, that they wanted preserved. In western public lands, like national parks and forests, fire officers would not enjoy the same latitude as their colleagues at Carolina Sandhills. The California model, not the Florida model, would prevail. Besides, worsening drought, continued construction along the wildland fringe, and a bottomless federal deficit have reduced the slack available for experimentation and errors. Rather than deliberate burning, federal fire officers will likely outsource more of the job to nature in the form of wildland fire use.

A different example of restoration comes from the Missouri Ozarks, where a middle ground is emerging: private lands are becoming public, or are emulating public lands, and administrators are creating working landscapes committed to public ecological goals rather than simple commodity production.

Here, there is little natural fire; the history of burning in the region closely follows the removal and subsequent reintroduction of people into the region. Because the rugged terrain provides readymade barriers to the spread of fire, people must fill up and bring fire to each parcel of the land. By the end of the 19th century, that process of settlement had saturated the land with fire but also built roads, introduced overgrazing and logging, and generally stripped the region of available fuels. The number of fires began to decline. As sections were then turned into national forests, active firefighting aided the process. Fires ceased to be a defining feature of the scene.

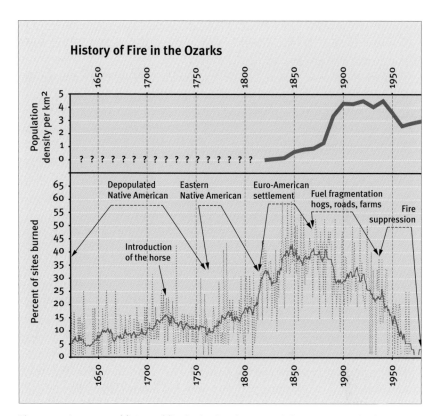

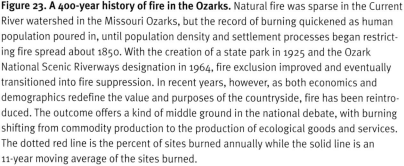

Figure 23. A 400-year history of fire in the Ozarks. Natural fire was sparse in the Current River watershed in the Missouri Ozarks, but the record of burning quickened as human population poured in, until population density and settlement processes began restricting fire spread about 1850. With the creation of a state park in 1925 and the Ozark National Scenic Riverways designation in 1964, fire exclusion improved and eventually transitioned into fire suppression. In recent years, however, as both economics and demographics redefine the value and purposes of the countryside, fire has been reintroduced. The outcome offers a kind of middle ground in the national debate, with burning shifting from commodity production to the production of ecological goods and services. The dotted red line is the percent of sites burned annually while the solid line is an 11-year moving average of the sites burned.

Eventually, people began to leave the landscape because it could no longer support them. The forest commenced regrowing. Some sites became new federal lands under provisions for protecting scenic rivers. The Missouri public created a commission to buy up land and commit it to recreational uses and nature protection. In 1951 a private landowner, Leo Drey, began purchasing significant tracts of land to create the Pioneer Forest, an experiment in sustainable forestry;

ultimately it was ceded to the L-A-D Foundation, which continues to reconcile harvesting with other ecological benefits. In brief, the land changed hands, changed character, and changed its relationship to fire.

On national forests, Nature Conservancy holdings, and state parks, efforts commenced to reintroduce fire. There are few prohibitions on intervening to create conditions that will favor the kinds of fire that administrators desire. The task has proved difficult—much trickier than it seemed. Reintroducing lost fires is akin to reintroducing lost species. But public support exists, there is room for maneuvering, and the result is a kind of new Missouri compromise that is restoring fire not for wilderness or specialty species habitats but for working landscapes committed to general public purposes such as recreation, environmental amenities, and ecological goods and services overall.

BIG FIRES, MEGAFIRES

The big fire has long haunted fire protection. Big fires do the most damage and ring up the greatest costs. The logic behind stopping every fire while small was to prevent some from becoming big. No program of fire protection could succeed, it was argued, if it tolerated big fires. Yet the fire revolution also has affected our understanding of big fires, for good and ill.

One revelation is that some big fires are both inevitable and necessary. Lodgepole pine, coastal Douglas-fir, southern California chaparral—all burn with high-intensity outbursts that become expansive and expensive to battle. Removing all such outbreaks, however, will stagnate the forest, and the effort is in any event doomed. The future points to wildland fire use, in which nature can make the necessary reckoning, and to prescribed crown fires that segregate intense burning from large-area burning. No program of fire management can succeed, it now seems, unless it tolerates some varieties of high-intensity fires.

At the same time, some big fires have mutated into monster burns on a scale not seen for decades, acquiring the label *megafires*. These enormous burns are savage in combustion intensity, complex in management, and dangerous to firefighters, and they seem to many agencies to be the wildfires of the future. In explaining their sudden appearance, the authorities, both political and scientific, have blamed severe climate and the unsavory legacy of past practices, notably fire suppression. The first culprit absolves the agencies from their inability to contain the outbreaks, and the second reconfirms a commitment to fire's more benign reintroduction. A more widely traveled mind, however, might note that megaburns have become a feature of recent decades around the globe. They have simmered in Amazonia and Borneo, plagued Spain and Portugal, and spread across Siberia

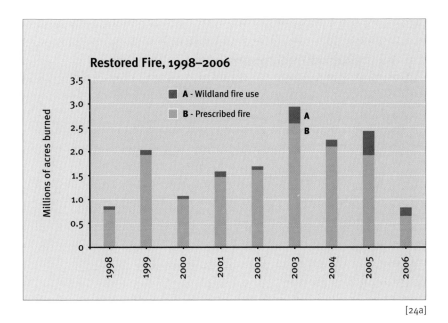

[24a]

Figure 24. Restored fire versus wildfire. Statistics for reintroduced fire date only from 1998. Federal agencies now employ both prescribed burns and wildland fire use (a), but the resulting area burned remains much less than the area burned by wildfire (b). The distribution of fires larger than 100,000 acres (c) corresponds not only with public lands but also with those states that have extensive contiguous wildlands; three states account for 67 percent of the burning.

and Mongolia. Their story involves far more than the parochial scene in the American West.

An explanation might look to three factors: climate, land use, and fire practices. The climatic component is obvious, a precondition for burning. The land-use factor encompasses any change in the condition of fuels over wide areas. This might mean logging, clearing land for oil palm plantations and pastures, draining wetlands, colonizing rural landscapes by houses, or reclassifying lands from production to protection. Wholesale land abandonment in Portugal is fueling some of the worst fires on the planet. But huge fires have broken out in Australia because land has been transferred from state forests to national parks and in the name of "naturalness" has sprouted thickets of combustible material. Climate alone is insufficient to account for the scale and location of such fires.

The third factor may be more surprising. Changes in how people relate to fire—how they respond to wildfire, whether they are inclined to do prescribed burning, how willing they are to risk firefighters by crowding firelines—also

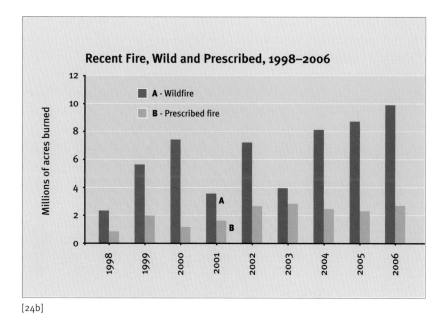

[24b]

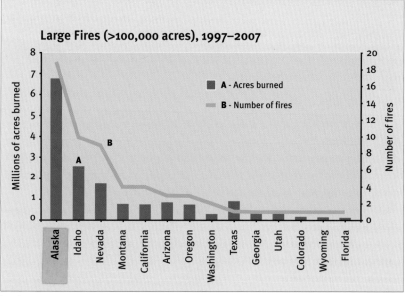

[24c]

affect the area burned. Studies that have tracked climatic factors over the past 35 years in the United States have pointed to earlier springs and drier winters, all of which can support more and larger fires. But burned area is not an index of climate alone; it is a proxy for climate interacting with people, for this same period of time coincides with the fire revolution, during which the federal agencies sought mightily to *increase* the area burned. The figures that show the public domain on fire also suggest that the agencies have succeeded. They have changed suppression tactics, promoted wildland fire use, and introduced prescribed burns, all of which have expanded the area blackened by fire. Be careful what you wish for.

Put differently, all the factors that propel megafires revolve around one agent: humans. People account for the changes in land use, they decide fire policy and practice, and they appear to be underwriting climate change through their combustion habits. The idea that physical factors—climate, fuels—alone explain the surge of megafires only deflects attention from the real culprit, ourselves. Any remedies to contain such monsters must consider what we do on the land and how we understand our stewardship over fire.

GLOBAL CHANGE, GLOBAL CHALLENGES

What happens in America affects other lands, and what happens on other lands affects America. Global climate change has global origins. Invasive grasses that thrive on fire from Central Asia (like *Bromus*) are rewriting the fire regimes throughout the American West and complicating attempts to reintroduce fire. The proposal to slash and burn tens of millions of acres of America's public lands in the name of ecosystem health, as proposed in the 2003 Healthy Forests Restoration Act, must explain why this is good in the United States but wrong if done in Indonesia. Why is it vital that Americans burn prairie preserves but essential that Africans stop firing savannas? Why is burning old-growth lodgepole pine an ecological imperative but burning old-growth elsewhere is a catastrophe for planetary biodiversity? How will the United States restore and fix the carbon liberated by its doctrine of fire restoration? How does American fire management propose to become carbon neutral?

Such issues may escalate into intellectual and political fire fronts as compelling as those around megafires. To date, Americans have hardly begun to factor such considerations into the calculations to get fire back on the land. But this unilateralism will not long prevail: the United States will have to explain its fire doctrines and compensate for the behavior of its fires. The argument is likely to hinge, again, on the inextinguishable value of fire's ecological presence. Fire is necessary because it does biological work essential to maintain ecological integrity in a desired landscape. If all that fire does is reduce fuel, there are other means available, some of which can also store carbon, as fire does not. These questions will, sooner rather than later, affect American fire practices and fire regimes.

The other side of global fire management is the emergence of institutions that cross borders. The founding era of fire protection had a global reach because empires connected colonies with common institutions and philosophies and foresters' guilds established communication across national boundaries. The recent era of decolonization has erased those bonds, even as it has confirmed the value, and perhaps necessity, of replacing them.

One useful institution is the International Forestry branch of the U.S. Forest Service, which promotes research and emergency aid. Another consists of bilateral treaties to assist with fire crises (the United States has such accords with Canada, Mexico, Russia, and Australia). Another relies on the United Nations through such agencies as the Food and Agriculture Organization and the International Strategy for Disaster Reduction, which helps sponsor the Global Fire Monitoring Center in Freiburg, Germany, the most robust single source of information about fire on Earth today. Under the auspices of the latter, a global fire network is emerging to create a truly planetary community of fire scientists and practitioners. To such endeavors might also be added the instinctively transnational interests of most researchers. More and more, fire modeling is becoming a collaborative enterprise. Fire science no more respects borders than does smoke.

Think global, act local—the details will continue to evolve over the next few decades, but the likely outcome is that effective fire management will be ever more site-specific even as our understanding of fire and our justification for it will assume planetary dimensions.

POSSIBLE FUTURES

Many fires in many places, many kinds of fire problems that require many kinds of site-specific solutions—that, perhaps, is the lesson learned from America's fire history. Fire is a creature of context: to control fire you need to control its context. The effort to impose a universal definition of fire and a universal standard of treatment—to make fire a government monopoly, to lodge it within one academic discipline and one profession—has failed. Instead, the reformation that boiled out of the 1960s has reestablished a principle of multiple solutions and sparked institutional reforms to allow them. This leaves a paradox. Even as fire seems more prominent in public consciousness, it is receding from daily lives and everyday landscapes. Industrial combustion has crowded out burning to restricted sites, most prominently nature preserves.

Amid the variety of those fires, three stand out as objects of contemporary concern: the wildland-urban interface, prescribed fire as a means of landscape restoration, and wildland fire use as a technique by which to allow fire a more robust role on natural landscapes. Each has its strengths, its liabilities, and its paradoxes.

In the interface. Even as the wildland-urban interface appeared likely to spread everywhere, the housing market began to collapse, the country seemed overbuilt in its suburbs, and the exurban landscape was reluctantly making the reforms required. The worst scenes are those communities developed before awareness became common; these will be difficult to retrofit and will likely become the scene of major damages in the near future. Still, the spectacular outbreaks are more or less localized into California.

The two areas that need further action are, first, to strengthen codes to make the house and its immediate surroundings (within a few feet) less receptive to flame and embers, and, second, to replace the increasingly mindless appeal to mass evacuations. The Australian experience is highly relevant: leave early or stay and defend. In reality there is no other option. Public agencies cannot hope to protect each house, and guidelines for defensible space are useless against firebrands, only one of which can take out an entire structure. Local people will have to defend local houses.

Prescribed burning. Deliberate fires will likely reflect the founding, bicoastal divide of the fire revolution. It will remain strongest where it has always flourished most: the Southeast, with Florida as an exporter of exemplars and enthusiasm. It will thrive best on working landscapes rather than in wild ones.

Regulations will become more restrictive and push burning away from fuel reduction as a dominant concern and pull it toward ecological work, which alone can justify the collateral damages from smoke, escapes, and errors. The costs are not trivial and will rise as smoke affects highways, the public health of nearby communities, and air quality generally, and as escape fires threaten adjacent lands and towns. Global climate issues will almost certainly bring prescribed fire under a regime of carbon accounting. Unless liability laws allow for more flexibility, the cost to individual practitioners and agencies may become prohibitive.

Yet history's lessons go beyond the tightening competition between open burning and industrial combustion. Prescribed burning too often follows a bureaucratic formula that establishes complex checklists, dates, particular sites; this is not how people in the past managed to burn immense landscapes, patch by patch. Aboriginal burners were opportunistic, nimble, able to forage for fuels ready to burn, and they repeated their practices over long centuries. One-time set-piece burns don't restore this system. Single burns don't establish new fire regimes.

Once fire has been removed, it is not easy to reintroduce. Grassy landscapes can rebuild more easily than woody ones, but prescribed fire may require other practices so that fire has a habitat that allows it to do what land managers wish. Recovering historical practices might allow this to happen on the ground. What will not help are assertions that logging can substitute for burning. The two processes are not interchangeable. Logging is a mechanical process that physically removes biomass; fire is a chemical process that changes biomass. Logging takes the big particles and leaves the small; fire burns the small and leaves the big. There is a case for landscape thinning—a kind of woody weeding. But the assertion that logging can act as a surrogate for burning has no legitimacy.

Wildland fire use. Current strategy will likely endure, although it will probably disappear as a term (the changing labels have proved unstable for decades). The approach may well become dominant on truly wild lands. It poses two special paradoxes.

First, wildland fire use demands close tending. This may take the form of dense information about landscapes and meticulous monitoring, not only for fires that are blowing and going but also for their ecological consequences. They may enhance the wild; they may not enhance biodiversity, ecological integrity, and so on. Moreover, the practice will not be cheap, easy, or safe. A small number of escapes and major smoke episodes may put the entire strategy under political scrutiny. The practice will flourish best in large blocks of land distant from urban centers.

Second, it demonstrates the interdependence of fire and land. The usual formula is that fire management should follow from land management. The purposes of the land will dictate what kinds of fires should occur in what way. But wildland fire use shows that the reverse is equally true. What kinds of fires a landowner allows will shape what kind of landscape exists. Committing a site to wildland fire use is committing it to wilderness. That the new federal policies, particularly those of the Forest Service, to say nothing of the dismal budgetary state of the nation, will promote wildland fire use or its equivalents will commit the country to an extraordinary experiment in wholesale ecological change. Amid climate change and invasive species, no one can say what the outcome will be.

Amerca's many fires have many futures. What might they be, and how might we assist in promoting those we desire? Consider the three narratives that, braided together, shape the story of America's fires.

Begin with the largest of fire's narratives, the transition to industrialization. Not only is it unlikely to end, it is accelerating as China, India, and other countries rapidly modernize, with consequent environmental effects. China's carbon emissions, for example, now exceed those of the United States. The likely outcome is greater pressure to reduce open burning of all varieties and to find technological substitutes that do not rely on internal combustion to power machines. These efforts will take decades at best. But they will focus the argument for prescribed fire, or for tolerating natural burns on wildlands, to the ecological role fire plays.

Consider, next, the second narrative, free-burning fire on public wildlands. These lands have become fire's primary habitat. What happens to them will shape the future of fire, just as our choice of fire practices will shape the character of those lands. Although private lands, such as those of The Nature Conservancy, are supplementing the public domain, the geography of fire will likely track the disposition of public lands and those private landscapes that are undergoing alteration—for example, timber holdings or agricultural lands converted to exurbs. Each is a source of wildfire. But whereas the public wildlands promise to endure for some time, the conversion of private lands will last only until a new equilibrium results, whose probable outcome will be less burning in the landscape.

And finally ponder the national narrative of institutions, ideas, and practices. Each element has predictable outcomes if trends persist. And each offers prospects to direct those movements.

Figure 25. The pyric transition: Two realms of combustion. The two extremes: Europe, aglow with electric lights, and sub-Saharan Africa, whose interior is aflame with open burning. Some of Europe has burned in recent years, notably Portugal and northwestern Spain, where economic conditions have sparked a wholesale exodus of the rural population, leaving the land fluff with combustibles, and also the Balkans, subject to abandoned lands and incendiary warfare. So, too, African metropolitan areas blaze with lights. The two realms tend not to coexist, or do so only temporarily.

FIRE INSTITUTIONS

By the early years of the 21st century, the fire revolution that bubbled up during the 1960s had rebuilt the institutional framework for modern fire management, yet in ways that, paradoxically, have left the fire community less certain how to proceed.

The final reforms came in a surge. In 1995 a common federal fire policy was approved that effectively ended the last echoes of the old suppression mandate. In 1998 Congress allocated significant funding for fire research among all agencies with the Joint Fire Science Program. In 2000, amid unstoppable wildfires in the Northern Rockies and two breakaway prescribed burns in the Southwest, the National Fire Plan gave political attention and heft to wildland fire management. In 2001 federal wildland fire policy was updated; the Wildland Fire Leadership Council was established to coordinate among the federal land agencies, supplementing the National Interagency Fire Center and National Wildfire Coordinating Group; and the Western Governors Association proposed "A Collaborative Approach for Reducing Wildland Fire Risks to Communities and the Environment: A 10-Year Comprehensive Strategy," effectively upgrading the federal-state cooperative programs that had begun 90 years before with the Weeks Act. In 2003 the Healthy Forests Restoration Act expanded the range of federal interventions to encourage thinning and burning.

Other institutions sought to update private and public-private fire practices. Building on traditions of regional fire councils originally devised for wildfire, prescribed fire councils sprang up, initially in Florida, and then throughout the Southeast and elsewhere, to strengthen and coordinate deliberate burning. The Nature Conservancy not only enlarged its programs but also accepted a role in brokering among other institutions by overseeing a web of "fire learning networks." More and more local authorities established codes for housing developments in fire-prone landscapes, and those communities organized for protection and created fire districts. A further blurring of boundaries has come from hiring private contractors to do what, 30 years ago, were government jobs; this extends even to prescribed burning. With a 20-year full-retirement option, former fire officers are becoming consultants and suppliers of services.

In 2008 the federal policy was reconfirmed and broadened by making "appropriate management response" the universal standard for administration. The old distinctions between wildfire, prescribed fire, wildland fire use, and so on all vanished, at least in principle. There was only fire and what agencies might choose to do about it.

This freedom comes at a cost. Fire officers no longer unthinkingly suppress every fire, but it is not clear just what they should do about fires that start by accident, nature, or arson, or what fires they might themselves introduce.

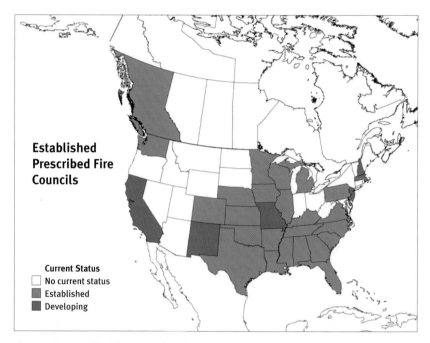

Figure 26. Prescribed fire councils. Like their predecessors, the timber protective associations, these regional prescribed fire organizations are part of the new geography of fire institutions. They have generally appeared where federal land is not the dominant ownership type.

Theoretically, the guidelines for action are embedded in detailed land management plans, but the fire program will be only as good as local skills and the local knowledge of place. It also places more burdens on fire officers, who are also urged to consider personal liability insurance and are otherwise made insecure. The new freedom makes more inventive, riskier behavior possible, yet holds individuals more accountable for failures.

FIRE IDEAS

Perhaps more unsettled is fire's intellectual standing. The revolution had begun with ideas, but those ideas had focused on fire's "naturalness" and on the contradictions of its suppression. Their proponents sought to reform practice and reintroduce fire. But the understanding of fire as a phenomenon is less easily reformed than the institutional order.

Today, fire seems to intersect more and more topics under the general rubric of global change. A study of fire-themed publications points to an astonishing

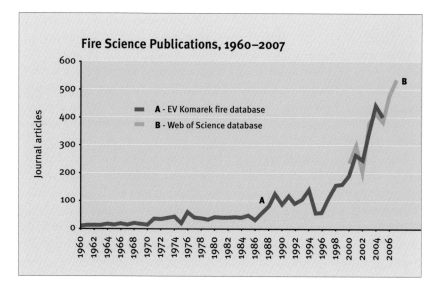

Figure 27. Fire science research. An outpouring of papers in research journals occurred after the policy of total fire suppression was abandoned. It was not science that caused a change in perception but a change of values, which then led to the sponsorship of more science to help guide practice. The growth during the late 1990s and early 2000s reflects, in part, new funding from the Joint Fire Science Program, plus the emergence of fire science in other parts of the world, and the global climate change agenda—particularly concerns over atmospheric chemistry and tropical biodiversity.

rise in recent years. In the 1960s, as America's fire revolution began, some 13 papers a year were published, about 30 percent of which concerned management. By 2000–2005 more than 300 papers were published, of which only five percent were directly about management. In journals published by Blackwell, Elsevier, and Springer, the number of fire articles has risen from an average of two per year in the 1970s to 75 per year since 2000. Accordingly, fire science, like most active fields, is fragmenting, yet, unlike comparable endeavors, it has no intellectual core. Fire remains something taught within the context of other disciplines; the only fire department at a university is the one that sends emergency vehicles when an alarm sounds. Instead, fire science remains overwhelmingly government-sponsored science, funded to answer questions regarding fire on the public domain. And despite proliferating studies, it does not really address the underlying concerns.

The prevailing physical paradigm holds that fire is a chemical reaction, the rapid oxidation of hydrocarbons, shaped by the characteristics of its physical environment. These determine how the zone of combustion moves about the landscape. Fire ecologists study how this physical process affects the living world. Fire management means identifying physical means of control, primarily through kindling, quenching, slashing, and otherwise shoving hydrocarbons around. The origins of this paradigm lie in the need of governments to control fire, for which they sought practical help from applied science. It told them how fast fire might spread, how intensely it might burn, and hence how difficult would be their task to control it. Besides, counting, measuring, and writing formulas looks like real science.

Yet it is equally possible to imagine fire as biologically constructed. Life creates fire's essentials: fuel, oxygen, and through people, ignition. Physical parameters like terrain, wind, and drought matter only insofar as they influence the biomass through which combustion spreads. We often say of a disease that it spreads like wildfire; it might be more apt to say of a fire that it spreads like a disease. Picturing fire as a contagion of combustion invites us to explore measures of biological control and allows us to address through common language and concepts those issues that most plague fire's management beyond the need to contain a flaming front—issues such as ecological integrity, sustainability, biodiversity, and environmental ethics. It transforms landscapes from fuel arrays into fire habitats. Not least, defining fire biologically would allow humans to find their own niche in fire's ecology.

Such a leap invites a third conception of fire: that fire is primarily now a cultural phenomenon. We have become the principal means for shaping fire regimes and are apparently even warping climate. Studied apart from human practices, fire is meaningless; our problems with fire, whether we have too much or too little, whether a million-acre burn is a disaster or an ecological marvel—are culturally conditioned. Even what we choose to study and how we choose to study it are a cultural call. The integration of the physical and biological parameters occurs, that is, within institutions and the realm of ideas.

What links fires kindled by lightning bolts on nature reserves with internal combustion in power plants is us. We are, after all, uniquely fire creatures with a species monopoly over fire's manipulation. Other creatures dig holes, knock over trees, hunt, eat plants: only we do fire. Our stewardship over fire is the signature of our unique ecological agency. Until we openly acknowledge our firepower, we cannot effectively exercise that stewardship. A cultural paradigm thus enjoins us to look for fire's management through cultural controls.

Each conception—physical, biological, cultural—can explain the whole of fire's phenomena on its own terms. Each can absorb the others. And each will address major topics differently. By way of illustration, consider how each addresses the megafires that have blistered the planet over the past 15 years. The physical paradigm might liken those fires to a tsunami, a geophysical force over which humanity has scant control. Physical countermeasures, if suppression and containment don't work, mean moving out of its way. The biological paradigm might, instead, liken those outbreaks to an emergent plague like avian flu, the outcome of broken biotas, favorable climate, and human habits that can propel a lethal pandemic. For appropriate countermeasures it might look to models from epidemiology and try to contain fire like a disease. The cultural paradigm would note that almost all of these disturbances have resulted from human land use or changes in institutions and liken megafires in the ecosystem to a mass revolt of the peasants, an ecological insurgency. Its countermeasures would start with modifying human behavior and understanding, which is to say, how we define and respond to the problem.

Each view brings special insights. The physical paradigm can instruct us in fire's control along its flaming front; the biological paradigm, the management of fire in landscapes and macrobiotic settings; the cultural paradigm, the role that humans should exercise, how and where we should apply our firepower. Choosing among them resembles a game of rock-scissors-paper. But at least today, the physical conception rules them all. Until the other fire paradigms acquire comparable intellectual standing, we will continue to do more of the same even when the results do little more than invite more research to continue the cycle. The revolution in ideas has barely begun.

THEN AND NOW

A century has passed since the Transfer Act of 1905 moved the U.S. forest reserves from the Department of the Interior to the Department of Agriculture and the modern era of wildland fire management commenced. Then and now—the eras are eerily symmetrical, only inverted. In 1905 the U.S. Forest Service confronted rapacious logging, reckless mining, damaged watersheds, boundary threats in the form of agricultural encroachment—and fire. Today it confronts sick forests, endangered species, invasive species, illegal logging, boundary threats in the form of urban encroachment—and fire.

In 1905 all sides exploited the fire scene to animate their messages, for nothing else mattered until we controlled fire. The perception was that the nation had too many bad burns, that the solution was to abolish all fires, and that the public could comprehend only a much-simplified message. There was vigorous dissent, however, over what fire management meant, whether, in particular, it should be based on firefighting or fire lighting. The great achievement of the young Forest Service was systematic fire protection and a story to sustain it.

Now all sides are again exploiting the fire scene to animate their messages, for nothing else matters until we have fire properly in hand. The perception among the fire community is that the nation has too few good burns, that the solution is to reinstate fire wherever possible, and that citizens cannot understand the complexities of the issues. This time, however, there is little dissent over fire management philosophies—this may be the most striking and troubling difference. The great achievement of this era of reformation is surely its determination to reconnect fire with land management, at least in principle.

Then, fire protection was part of a global undertaking in state-sponsored conservation. Now, state-sponsored forestry has run its course. The old model has imploded: relict institutions have been scrapped or retrofitted with new inner workings better suited to the goals of an urban and industrial society. Today, fire management must again situate itself within a global context. It does so directly in the case of global warming, which is at its base a question of combustion. It does so indirectly for problems of land conversion, nature preservation, biodiversity, and the like, where fire has been inserted into places that can't accept it and withheld from places that need it.

Then, big fires kindled a sense of crisis. They were visible emblems of a nature knocked out of balance and careening into chaos. Looming over all stood the Big Blowup of 1910. Today, megafires have returned, again as powerful tokens of a nature gone awry. But the giant smokes rising over the Northern Rockies in 2000 or the flames riding a shattering wind into the fringes of San Diego in 2003 are not the Big Burn of today. For that, look not to towering convective columns of free-burning flame but the often invisible spumes of industrial combustion.

In the future fire will continue to compete with industrial combustion. It will morph as changing climate and altered land uses deliver or withdraw suitable fuels. It will depend on the character, or even the survival, of public lands. It will change with the various institutions that society commits to its oversight. But it will also join a global forum across peoples, institutions, and values, in which what one nation or professional group does must answer to other nations and groups; an epoch of unilateral fire management may draw to a close. A century ago foresters controlled the global agenda of fire management. Today they do not. Yet the founding questions endure: how much can we apply and withhold fire, and how and why should we try to do so.

The raw rangers of a fledgling Forest Service understood fire's primacy only too well in 1905. They struggled mightily to control those flames. Even before the Big Blowup, Henry Graves declared that fire protection was 90 percent of American forestry, and his successors continued to obsess over fire into the postwar era. By our reckonings, those early rangers got fire wrong; they let abusive fire override all other concerns.

As the summer 2007 fire season once again roared over the Northern Rockies, five former Forest Service chiefs signed a public statement that fire management was threatening to overwhelm everything else the agency wished to do. Fire costs had become a black hole. In 1990 fire management claimed less than 15 percent of the agency's budget; by 2006 that figure had swollen to 45 percent; by summer 2007, it was more than 50 percent and rising. Chief Dale Bosworth elaborated that "if you look down the line three, five, ten years from now, firefighting costs could presumably be close to 100 percent." Fire management had come a full, vicious cycle, as though a century had made not a whit of difference.

But of course it had, and the true lesson was not this policy or that, this prescription or that one, but exactly what those early rangers understood only too well: fire mattered. The real error had been to imagine fire as a one-time task and a precondition to true management, instead of the dominant force on the scene. The potential error today is again to define fire apart from everything else, to suggest that it can be isolated, managed, harnessed to budgets and five-year plans, when all the evidence of a century's intense encounter says otherwise—says that it is consuming the agency's budget precisely because it cannot be segregated from land management. Fire's return is based on a misperception; it had never left. It had been there all along, as fundamental to wildlands as sun and rain, but only—for a while—suppressed. It has reclaimed the agency's mission because it has always been there.

Fire matters because nothing else can so alter the land at a stroke, which is why fire remains today, as at the founding of forestry in the United States, the great instrument of any land stewardship that seeks to preserve or enhance the ecological integrity of a place, and why, a century from now, scholars will look back and say that we too got it wrong in our particulars. What we can hope is that we will get the basics right: that fire is something many lands need, and that we as a species must do our part to get it right with dedication, humility, and tolerance.

SELECTED READING

Arno, Stephen F., and Carl E. Fiedler. 2005. *Mimicking Nature's Fire: Restoring Fire-Prone Forests in the West*. Washington, DC: Island Press.

Boyd, Robert, ed. 1999. *Fire, Indians, and the Land in the Pacific Northwest*. Corvallis: Oregon State University Press.

Carle, David. 2002. *Burning Questions: America's Fight with Nature's Fire*. New York: Praeger Publishers.

Crutzen, P. J., and J. G. Goldammer, eds. 1993. *Fire in the Environment: The Ecological, Atmospheric, and Climatic Importance of Vegetation Fires*. New York: John Wiley & Sons.

Delcourt, Paul, and Hazel Delcourt. 2004. *Prehistoric Native Americans and Ecological Change: Human Ecosystems in Eastern North America since the Pleistocene*. Cambridge: Cambridge University Press.

Friederici, Peter, ed. 2003. *Ecological Restoration of Southwestern Ponderosa Pine Forests*. Washington, DC: Island Press.

Goudsblom, Johan. 1992. *Fire and Civilization*. London: Penguin Press.

Guyette, R. P., R. Muzika, and D. C. Dey. 2002. "Dynamics of an anthropogenic fire regime," *Ecosystems* 5: 472–86.

Kennedy, Roger G. 2006. *Wildfire and Americans*. New York: Hill and Wang.

Lewis, Henry T., and Teresa M. Ferguson. 1988. "Yards, corridors, and mosaics: How to burn a boreal forest," *Human Ecology* 16: 57–77.

Omi, Philip. 2005. *Forest Fire: A Reference Handbook*. Santa Barbara, CA: ABC Clio.

Pinchot, Gifford. 1905. *The Use of the National Forest Reserves*. Washington, DC: Government Printing Office.

Pyne, Stephen J. 1995. *Fire in America: A Cultural History of Wildland and Rural Fire*. Seattle: University of Washington Press, 1995.

———. 2001. *Fire: A Brief History*. Seattle: University of Washington Press.

————. 2001. *Year of the Fires: The Story of the Great Fires of 1910*. New York: Viking.

————. 2005. *Tending Fire: Coping with America's Wildland Fires*. Washington, DC: Island Press.

Pyne, Stephen J., Patricia L. Andrews, and Richard D. Laven. 1996. *Introduction to Wildland Fire*, 2nd ed. New York: John Wiley & Sons.

Report of the National Commission on Wildfire Disaster 1994. Washington, DC: Government Printing Office.

Rothman, Hal K. 2007. *Blazing Heritage. A History of Wildland Fire in the National Parks*. New York: Oxford University Press.

Sauer, Carl. 1967. *The Early Spanish Main*. Berkeley: University of California Press.

Schiff, Ashley. 1962. *Fire and Water: Scientific Heresy in the Forest Service*. Cambridge, MA: Harvard University Press.

Stewart, Omer C. 2002. *Forgotten Fires: Native Americans and the Transient Wilderness*. Norman: University of Oklahoma Press.

Tall Timbers Research Station. 1962–1997. *Proceedings, Tall Timbers Fire Ecology Conferences*, vols. 1–20. Tallahassee, FL: Tall Timbers Research Station.

U.S. Department of the Interior and Department of Agriculture. 1995. *Federal Wildland Fire Management: Policy and Program Review*. December 18.

Williams, Gerry W., comp. 2003. "References on the American Indian Use of Fire in Ecosystems." Unpublished study. Washington, DC: U.S. Forest Service, March 7.

Wright, Henry A., and Arthur W. Bailey. 1982. *Fire Ecology—United States and Southern Canada*. New York: John Wiley & Sons.

FIGURE SOURCES

Figure 1. (a) National Oceanic and Atmospheric Administration, http://www.lightning safety.noaa.gov/lightning_map.htm. (b) Mark J. Schroeder and Charles A. Buck, *Fire Weather*, Agriculture Handbook 360 (Government Printing Office, 1970).

Figure 2. H. W. Gabriel and F. Tande, "A regional approach to fire history in Alaska," *BLM-Alaska Technical Report* 9 (Anchorage: Bureau of Land Management, 1983).

Figure 3. Data from E. V. Komarek, "The natural history of lightning," *Proceedings, Tall Timbers Fire Ecology Conferences* 3: 139–84 (Tallahassee, FL: Tall Timbers Research Station, 1964).

Figure 4. (a) *Report on the Lands of the Arid Regions of the United States* (Government Printing Office, 1878). (b) Charles S. Sargent, *Forests, Report for the 1880 Census* (Government Printing Office, 1882).

Figure 5. (a) Map from U.S. Geological Survey, http://www.usgs.gov/hazards/wildfires/. (b) Data from D. E. Ward, et al., "An inventory of particulate matter and air toxic emissions from prescribed fires in the USA for 1989," *Proceedings of the Air and Waste Management Association 1993 Annual Meeting and Exhibition*, 93-MP-6.04 (1993).

Figure 6. (a) Forest History Society Photo Collection FHS2526. (b) Forest History Society Photo Collection R9_175688 (USFS Photo 175688).

Figure 7. (a) Forest History Society Photo Collection FHS2638. (b) Forest History Society Photo Collection FHS2524.

Figure 8. Data from S. T. Dana, *Forest and Range Policy in the United States* (McGraw-Hill, 1956).

Figure 9. (a) Forest History Society Photo Collection FHS2658 (USFS Negative 43822). (b) Forest History Society Photo Collection FHS2658 (USFS Negative 23991), photo taken by Henry Graves. (c) Forest History Society Photo Collection FHS5536 (USFS Negative 218443).

Figure 10. (a) Data from E. Pierce and W. Stahl, *Cooperative Forest Fire Control: A History from Its Origin and Development under the Weeks and Clarke-McNary Acts* (U.S. Forest Service, 1964). (b) Data from Elaine Kennedy Sutherland, Todd F. Hutchinson, and Daniel A. Yaussy, "Introduction, study area description, and experimental design," in Elaine Kennedy Sutherland and Todd F. Hutchinson, eds., *Characteristics of Mixed-Oak Forest Ecosystems in Southern Ohio Prior to the Reintroduction of Fire*, General Technical Report NE-299 (U.S. Forest Service, 2003). (c) Bureau of the Census, *Historical Statistics of the United States* (Government Printing Office, 1977).

Figure 11. (a) U.S. Forest Service, Historic Photo Collection, National Archives. (b) Forest History Society Photo Collection (USFS Negative 27297A) (c and d) U.S. Forest Service, Historic Photo Collection, National Archives.

Figure 12. (a) Forest History Society Photo Collection FHS3527 (USFS Negative 176440). (b) Forest History Society Photo Collection FHS3730 (USFS Negative 162629), photo taken in 1921. (c) Forest History Society Photo Collection (USFS Negative 391994).

Figure 13. (a) U.S. Forest Service, Historic Photo Collection, National Archives. (b) Forest History Society Photo Collection FHS1185 (USFS Negative 279337), photo taken at the Puerto la Cruz Camp, Aguanga Mountain, June 1933.

Figure 14. U.S. Forest Service, Cooperative Fire Program files.

Figure 15. (a) Forest History Society Photo Collection FHS3762 (USFS Negative 479458), specially staged photo taken August 10, 1955. (b) U.S. Forest Service, Intermountain Fire Sciences Lab.

Figure 16. (a) Data from Bureau of the Census, *Historical Statistics of the United States*. (b) Data from U.S. Forest Service. (c) Data from National Wildfire Coordinating Group, "Historical Wildland Firefighter Fatalities, 1910–1996," *National Fire Equipment System*, 1849 (March 1997), 2nd ed.

Figure 17. D. J. Novak, "Historical vegetation change in Oakland and its implications for urban forest management," *Journal of Arboriculture* 19(5) (1993): 313–19.

Figure 18. Florida Division of Forestry.

Figure 19. (a) Data from Laboratory of Tree-Ring Research, University of Arizona. (b) Data from U.S. Forest Service, Southwest Region.

Figure 20. Data from Garrett W. Meiggs, "Recent patterns of large fire events on Kaibab Plateau, Arizona, USA," Honors thesis, Cornell University (May 2004), and National Park Service.

Figure 21. Data from The Nature Conservancy.

Figure 22. (a) Data from Carolina Sandhills National Wildlife Refuge. (b) Data from National Interagency Fire Center.

Figure 23. Guyette, Richard P. and Bruce E. Cutter. "Fire History, Population, and Calcium Cycling in the Current River Watershed." In *Proceedings, 11th Central Hardwood Forest Conference*, Columbia, MO, March 23–26, 1997: 355–373. General Technical Report NC-188 (Washington, DC: U.S. Department of Agriculture, 1997).

Figure 24. Data from National Interagency Fire Center; historical fire statistics.

Figure 25. DMSP nighttime lights processed by the NOAA National Geophysical Data Center, courtesy Chris Elvidge.

Figure 26. National Coalition of Prescribed Fire Councils.

Figure 27. Data from Peter Frost, "Hot topics and burning issues: Fire as a driver of system processes—past, present, and future," paper presented to postgraduate course at C. T. deWit Graduate School for Production Ecology and Resource Conservation at Wageningen University, the Global Fire Monitoring Center/Max Planck Institute for Chemistry, and the United Nations University (2008).

ABOUT THE AUTHOR

Stephen J. Pyne is a professor in the Human Dimensions Faculty, School of Life Sciences, Arizona State University, a past president of the American Society for Environmental History, and the recipient of numerous fellowships and honors, including the Robert Kirsch Award for body-of-work contribution to American letters.

He has written a score of books, of which a dozen deal with fire on Earth. Among them are fire histories of the United States (*Fire in America*), Australia (*Burning Bush*), Canada (*Awful Splendour*), Europe including Russia (*Vestal Fire*), and Earth overall (*World Fire*; *Fire: A Brief History*). Others include two editions of *Introduction to Wildland Fire* and two current-affairs studies, one for the U.S. (*Tending Fire*) and one for Australia (*The Still-Burning Bush*).

His interest in fire grew out of 18 seasons with the National Park Service, of which 15 were spent with the North Rim Longshots, the subject of his book, *Fire on the Rim*. He is presently contributing his part to the exurban fire problem with a cabin bordering the Apache National Forest. He confesses to having started and stopped a fire on every continent.